50 Advances in Polymer Science

Fortschritte der Hochpolymeren-Forschung

W0107029

Editors: H.-J. Cantow, Freiburg i. Br. · G. Dall'Asta, Colleferro · K. Dušek,
Prague · J. D. Ferry, Madison · H. Fujita, Osaka · M. Gordon, Colchester
J. P. Kennedy, Akron · W. Kern, Mainz · S. Okamura, Kyoto
C. G. Overberger, Ann Arbor · T. Saegusa, Kyoto · G. V. Schulz, Mainz
W. P. Slichter, Murray Hill · J. K. Stille, Fort Collins

Unusual Properties
of New Polymers

With Contributions by
J. Pitha, G. Smets and D. Wöhrle

With 32 Figures

Springer-Verlag Berlin Heidelberg GmbH 1983

Editors

Prof. Hans-Joachim Cantow, Institut für Makromolekulare Chemie der Universität, Stefan-Meier-Str. 31, 7800 Freiburg i. Br., BRD

Prof. Gino Dall'Asta, SNIA VISCOSA – Centro Studi Chimico, Colleferro (Roma), Italia

Prof. Karel Dušek, Institute of Macromolecular Chemistry, Czechoslovak Academy of Sciences, 162 06 Prague 616, ČSSR

Prof. John D. Ferry, Department of Chemistry, The University of Wisconsin, Madison, Wisconsin 53706, U.S.A.

Prof. Hiroshi Fujita, Department of Macromolecular Science, Osaka University, Toyonaka, Osaka, Japan

Prof. Manfred Gordon, Department of Chemistry, University of Essex, Wivenhoe Park, Colchester C04 3 SQ, England

Prof. Joseph P. Kennedy, Institute of Polymer Science, The University of Akron, Akron, Ohio 44325, U.S.A.

Prof. Werner Kern, Institut für Organische Chemie der Universität, 6500 Mainz, BRD

Prof. Seizo Okamura, No. 24, Minami-Goshomachi, Okazaki, Sakyo-Ku, Kyoto 606, Japan

Prof. Charles G. Overberger, Department of Chemistry, The University of Michigan, Ann Arbor, Michigan 48 104, U.S.A.

Prof. Takeo Saegusa, Department of Synthetic Chemistry, Faculty of Engineering, Kyoto University, Kyoto, Japan

Prof. Günter Victor Schulz, Institut für Physikalische Chemie der Universität, 6500 Mainz, BRD

Dr. William P. Slichter, Chemical Physics Research Department, Bell Telephone Laboratories, Murray Hill, New Jersey 07 971, U.S.A.

Prof. John K. Stille, Department of Chemistry, Colorado State University, Fort Collins, Colorado 805 23, U.S.A.

ISBN 978-3-662-15320-8   ISBN 978-3-540-39515-7 (eBook)
DOI 10.1007/978-3-540-39515-7

Library of Congress Catalog Card Number 61-642

This work is subject to copyright. All rights are reserved, whether the whole or part of the material is concerned, specifically those of translation, reprinting, re-use of illustrations, broadcasting, reproduction by photocopying machine or similar means, and storage in data banks. Under § 54 of the German Copyright Law where copies are made for other than private use, a fee is payable to the publisher, the amount to "Verwertungsgesellschaft Wort", Munich.

© Springer-Verlag Berlin Heidelberg 1983
Originally published by Springer-Verlag Berlin Heidelberg New York in 1983.
Softcover reprint of the hardcover 1st edition 1983

The use of general descriptive names, trademarks, etc. in this publication, even if the former are not especially identified, is not to be taken as a sign that such names, as understood by the Trade Marks and Merchandise Marks Act, may accordingly be used freely by anyone.

Typesetting and printing: Schwetzinger Verlagsdruckerei.
2152/3140 – 5 4 3 2 1 0

Table of Contents

Physiological Activities of Synthetic Analogs of Polynucleotides

Josef Pitha

Macromolecular Chemistry Section, National Institute on Aging-GRC, National Institutes of Health, Baltimore, Maryland 21224, USA

The term polynucleotide is usually used for an analog or a fragment of nucleic acids. As with other important natural compounds, more distant analogs of polynucleotides, were synthesized and studied. In this review polymers which have backbones analogous to those of plastics and substituents analogous to those of polynucleotides are described; further in the text these compounds are named polynucleotide analogs. The interactions of polynucleotide analogs with natural polynucleotides and related proteins are described. These interactions strongly depend on the electric charge of polynucleotide analogs. Electroneutral analogs of polynucleotides interact with natural polynucleotides in a specific manner forming base-pair type complexes. On the other hand, the enzymes of nucleic acid synthesis are not directly bound by these polymers. Because neither polynucleotide analogs nor complexes of these analogs and natural polynucleotides can act as templates in biosynthesis, polynucleotide analogs can be used to block the natural ones and thus, to act as template-specific inhibitors of nucleic acid and protein biosynthesis. This inhibitory action of polynucleotide analogs is strong and specific in cell-free systems, and because they are not biodegradable it may be assumed that the effects of these polymers on cells or animals would also be strong and long-lasting. However, this is not the case; these effects were found to be rather weak and short-lasting. The observed decrease in effectiveness is the result of two factors: a) the ability of polymers to penetrate into the interior of cells is very low and b) by autophagy, the cells are able to capture foreign polymers, that penetrate their cytoplasm, in membrane-coated vesicles and thus, isolate these polymers from the processes occurring in the cell interior.

Advances in Polymer Science 50
© Springer-Verlag Berlin Heidelberg 1983

1 Introduction

In general, synthetic polymers do not possess strong pharmacological or physiological activities; consequently, the design and study of physiologically active polymers may be described either as a field of the future or a field of broken promises. This review was written with the intention to limit the unreasonable expectations that an uninitiated chemist may have about a newly prepared compound. Such limitations may help to establish which bioassays should be performed with a particular polymer and, in this manner, help with the eventual preparation of a polymeric drug.

Water soluble polymers are well represented in the human environment and in food. Thus, our very existence constitutes solid proof of the lack of the physiological effects of many of these compounds. Nevertheless, some water soluble synthetic polymers, even at very low concentrations, influence enzymatic processes that form the basis of the physiology of the body. The reason for a general lack of bioactivity of synthetic polymers on the organism's level is the inability of polymers to penetrate to the location where the body's basic biochemical processes occur. The human body's most prevailing component is water ($> 60\%$). However, this body of water is not a continuous phase, it is subdivided by lipid membranes into spaces of microscopic size. Lipids constitute about 15% of body weight and a considerable portion of that amount is used to form and maintain cellular membranes, a structural element of the body that diminishes the mobility of hydrophilic polymers in organisms.

To penetrate through an organism, a synthetic compound must not only be water soluble but also must be lipid soluble. Water solubility is required for diffusion through aqueous spaces and lipid solubility is necessary for penetration through lipid membranes, a process that occurs through phase-phase transfer. Experiences gathered in organic chemistry laboratories remind us that many compounds handled there have this characteristic; aqueous phase saturated with benzene has a rather strong smell, i.e., even benzene is reasonably well soluble in water. The situation in the field of polymers is different (e.g., polystyrene can be dissolved in a nonpolar solvent but does not detectably dissolve in water; poly-1-vinylpyrrolidinone is water soluble but any attempts to dissolve it in nonpolar solvents are bound to fail). Thus, in the field of polymers there are quite stringent limitations to their transport and distribution in the body. These limitations lead to considerable inertness of water soluble polymers on cell, tissue, and body levels and represent a serious obstacle for some applications. Nevertheless, these limitations do not prevent other applications and possibly they can even be used constructively; one can try to affect cells or organisms from the outside, i.e., by the surface reaction between polymer and biomembranes.

This review is divided into two parts:
1) the effects of polynucleotide analogs on enzymatic systems and
2) the effects of polynucleotide analogs on cell and oranism levels.

2 Chemical and Biophysical Backgrounds

Nucleic acids are carriers of genetic information and their importance has led to intensive interest in their structure and chemistry. Nucleic acids are formed from polymeric strands that have backbones composed from alternative phosphodiester and sugar units, and carry on every sugar unit a herocyclic substituent that is called a base (Fig. 1). There is one electronegative phosphodiester group per one unit of a strand an, consequently, there is strong coulombic repulsion between strands. Nevertheless, two or more strands may associate through the formation of hydrogen bonding between the heterocyclic bases; in this complexing adenine pairs with uracil or thymine (i.e., forms base pairs with these compounds), and guanine pairs with cytosine (Fig. 1). From a structural point of view, three aspects of nucleic acid structure seem to have paramount importance:
a) the stereoregularity,
b) the polyanionic character, and
c) the ability of strands to associate through the formation of specific hydrogen bonding between bases.

Further in this review we will deal with fully synthetic analogs of polynucleotides, (compare examples in Fig. 2.), that do not have *stereoregular* structures. Thus, the first important aspect of similarity to nucleic acids is lost. The absence of stereoregularity renders polynucleotide analogs ineffective in biological syntheses where natural polynuc-

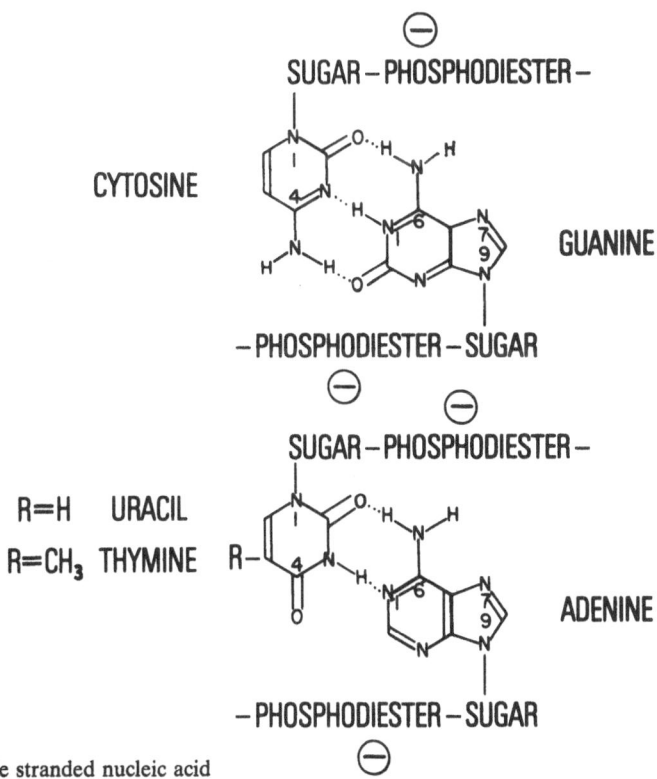

Fig. 1. Structure of double stranded nucleic acid

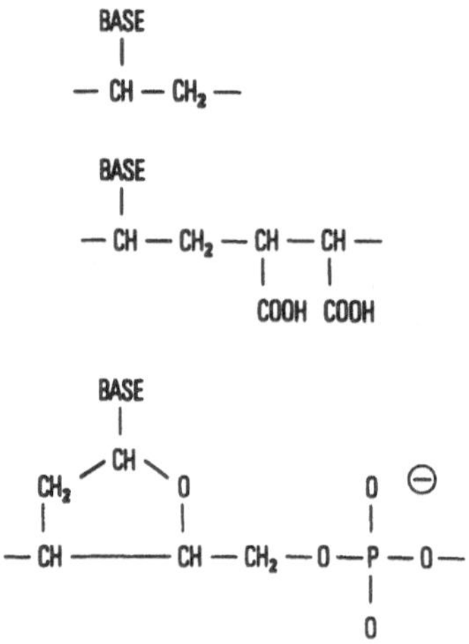

BASE
|
— CH — CH₂ —

BASE
|
— CH — CH₂ — CH — CH —
| |
COOH COOH

BASE
|
     CH
CH₂   O   O ⊖
| | |
— CH ——— CH — CH₂ — O — P — O —
|
O

Fig. 2. Structures of backbones of polynucleotide analogs and natural polynucleotides. If base in question is adenine, then the compounds are: 1) poly-9-vinyl-adenine, 2) copolymer of 9-vinyladenine and maleic acid, and 3) polydeoxyadenylate

leotides are required as templates. Apparently, the mutual fit between the enzymes (or ribosomes) and nucleic acid (polynucleotides) is so tight that the lack of stereoregularity cannot be tolerated. Nevertheless, polynucleotide analogs are good templates in nonenzymatic chemical syntheses of natural polynucleotides. In such chemical reactions the template serves only to increase local concentrations of the monomeric units and there are no requirements for a precise fit.

Electric charge has crucial importance to the bioeffects of polynucleotide analogs and can be used to subdivide these compounds into three subgroups (i.e., electroneutral, polyanions, polycations) that have distinctly different biochemical effects.

The analogs of the first subgroup do not have an electric charge (Fig. 2). This type of polynucleotide analog has been investigated earlier and more intensively than the others (e.g., the first monomers and polymers containing pyrimidine moiety were already prepared in the fifties[1]). Presently, many electroneutral monomers carrying various bases of nucleic acids are known and their polymerizations were systematically studied[44]. These polymers have a good ability to base pair with polynucleotides and with nucleic acids and they form several additional types of complexes that have not been observed in the polynucleotide field. This is a direct result of the absence of coulombic repulsions that, in the field of nucleic acids, can destabilize some weaker interactions. Whereas electroneutral polynucleotide analogs bind in a specific way to natural polynucleotides, the binding fo these polymers to enzymes of nucleic acid synthesis and degradation is weak. This is also a result of the absence of an electric charge; coulombic attractions are probably important for the strength of nucleic acid-enzyme binding. The formation of a complex between the natural polynucleotide and an electroneutral polynucleotide analog causes the natural polynucleotide to lose its ability to act as a template in biological syntheses.

Polynucleotide analogs that are polyanions (Fig. 2) have again a number of properties in common. The association of polyanionic analogs with natural polynucleotides is rather weak or, in some cases, even non existent. This is due to a combination of the presence of electronegative charges and the lack of stereoregularity. Thus, whereas there are as many coulombic repulsions in effect as in complexes of nucleic acids, only a few base pairs, which hold different strands together, can be formed. On the other hand, the presence of an electronegative charge leads to a strong association of these analogs with enzymes that bind to nucleic acids. Coulombic forces are probably of primary importance in these associations because the presence of bases seems to have only a secondary role.

Polycationic analogs have properties distinctly different from the previous subgroups. These analogs bind strongly to nucleic acids, but this is due to polycation-polyanion type interactions since the presence of bases has only secondary effects on these interactions; the templating ability of nucleic acids is blocked by polycationic analogs.

From various physical and biophysical properties of nucleic acid analogs the most important property for the present purpose is their interaction with nucleic acids. The spectrophotometric methods for detection of complex formation were applied to all combinations of polyvinyl polynucleotide analogs and natural polynucleotides (Fig. 3). In aqueous media hypochromic complexes were formed in combinations where the bases in the polynucleotide and analog were complementary. Poly-1-vinylcytosine is soluble in aqueous-propylene glycol; base-pair type complexes were detected there also. An analog of polyinosinate, poly-9-vinylhypoxanthine, is soluble only in solutions of a detergent, sodium dodecylsulfate. This detergent intercalates into the polymer and conveys to it an

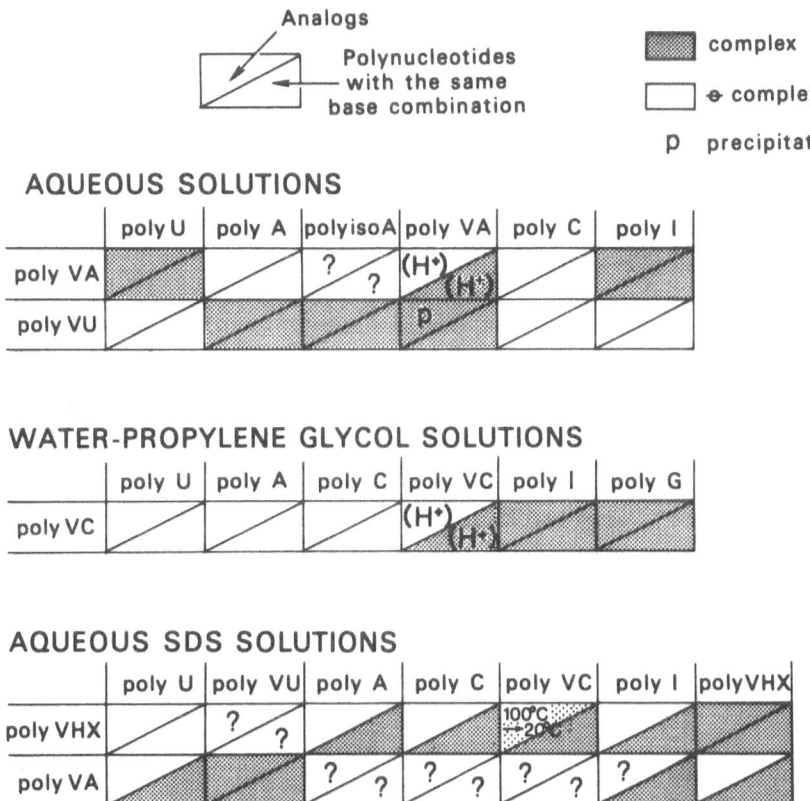

Fig. 3. Formation of complexes between various combinations of polynucleotide analogs and natural polynucleotides. Abbreviations used: poly VA, poly-9-vinyladenine; poly VU, poly-1-vinyluracil; poly VC, poly-1-vinylcytosine; poly VHX, poly-9-vinylhypoxanthine; SDS, sodium dodecyl sulfate

electronegative charge that considerably reduces the polymer's ability to bind to polyanions (Fig. 3). The same phenomenon was observed when poly-9-vinyladenine was studied in the same solvent (Fig. 3). The field of complex formation between various polynucleotides and their analogs has been reviewed several times[44].

The stoichiometry of complexes of polynucleotides with their electroneutral analogs was studied in detail[2–8]. The results can be summarized as follows:

a) when the analog contains a pyrimidine type base, i.e., uracil, cytosine, the complex contains several pyrimidine bases per one purine base of polynucleotide, i.e., adenine, guanine, hypoxanthine,

b) when the analog contains purine and the polynucleotide contains pyrimidine, the ratio of bases in the complex is close to one, and

c) the stoichiometry of the complex of poly-9-vinyladenine with polyinosiate, a complex where only purine type bases are present, also has a ratio of bases close to one[9, 10].

A rotary spectra of complexes suggests that there are some local structural regularities but, of course, the complexes are basically amorphous materials.

In addition to complexes that are detectable by spectophotometry, there are several complexes that are probably rather weak and can only be detected by methods that directly measure the binding between macromolecules[11].

3 Effects of Polynucleotide Analogs on Enzymatic Systems

Deoxyribonucleic acid (DNA), the genetic material of bacteria and eukaryotic cells, is either copied into new complementary DNA or transcribed into ribonucleic acid (RNA), also of complementary sequence. It is then used as a template in the so-called translation into proteins. Some viruses use RNA as genetic material that is copied into new complementary RNA, or in leukemie viruses transcribed into DNA; the latter process is called reverse transcription.

Numerous studies on the effect of analogs on these templated processes have been performed. From prokaryotic systems the RNA polymerase[12] and cell-free protein synthesizing systems[10, 13–16] were investigated; both of these were obtained from *E. coli*. From eukaryotic systems all three known DNA dependent DNA polymerases (alpha, beta, and gamma) were investigated; these were obtained from mouse tissues[17]. These polymerases are responsible for copying nuclear DNA. Polymerases that are responsible for the replication of mitochondrial DNA have not been tested. The effects of polynucleotide analogs on terminal deoxynucleotide transferase, a polynucleotide synthesizing enzyme from calf thymus, was also investigated[18]. The eukaryotic protein synthesizing system was prepared from Krebs II cells and the effects of polynucleotide analogs on this system was tested; polynucleotides or natural messenger RNAs were used as templates in these experiments[19]. In a majority of the experiments with viral enzymes[20–22], a crude preparation of reverse transcriptase from murine leukemia virus was used. However, several other experiments were performed using purified transcriptase from avian myeloblastosis virus to reconfirm the results[17, 21].

The general conclusions derived from these experiments are:

a) electroneutral analogs of polynucleotides cannot function as templates of enzymatic synthesis,

b) polynucleotides that are templates of enzymatic syntheses completely lose this ability after complexation with the complementary electroneutral analogs of polynucleotides.

In a majority of the experiments on the effects of polynucleotide analogs on templated biosynthesis, it is obvious whether or not the polynucleotide analog is capable of forming a complex with the template; nevertheless, sometimes the situation is complicated. Some polymerases require not only the template to which the complementary strand is synthesized. but also a primer, an oligonucleotide that is complementary to the template and which polymerase elongates. Here there is a possibility of blocking either template or primer, whichever is complementary to the analog. Whether or not such blocking occurs depends on the relative stability of various complexes. When polymerase uses a very stable template-primer complex, the electroneutral analog may have no effect even when it is complementary. Thus, when RNA polymerase transcribes polynucleotide in which adenine and thymine bases alternate and which forms a very stable double-stranded helix, no inhibition is observed with either poly-1-vinyluracil or poly-9-vinyladenine[12]. Nevertheless, these analogs effectively inhibit the very same enzyme when it uses single-stranded polynucleotide as a template[12].

There are two possible mechanisms by which a polynucleotide analog can inhibit the polymerase in the template-specific manner. Polynucleotide analogs can displace polymerase from its template and release the enzyme into solution in a free form; alternatively, the polymerase can become blocked on the template polynucleotide analog complex and remain there in a bound, inactive form. These alternatives were investigated for two enzymes, alpha class DNA polymerase and *E. coli* RNA polymerase. In both cases the polymerase becomes blocked on the polynucleotide templates rather than being released in free form[12, 17].

The effects of analogs on templated synthesis are selective, i.e., if there is no formation of a strong complex between the electroneutral polynucleotide analog and template-primer, there is no inhibition. Thus, electroneutral analogs are template-specific inhibitors and do not block polymerase activity in a nondiscriminatory way. This may have potential use in a number of situations, e.g., in combating viral infection it may be preferable to block just viral templates rather than blocking both the viral and host templates. It is important to realize tht natural polynucleotides and electronegative polynucleotide analogs inhibit polymerases mainly by binding directly to an enzyme, and this type of inhibition, of course, is not template-specific. Numerous examples of such binding-inhibition by polyanions were observed and have been reviewed[23]. For example, a carboxylate-type polymer, prepared by copolymerization of divinylether and maleic anhydride, is bound to viral reverse transcriptase and inhibits the enzyme[24]. A similar inhibition of viral reverse transcriptase was obtained with copolymers of either 1-vinyluracil or 9-vinyladenine with acrylic acid; inhibition in these systems occurred irrespective of complementarity[18, 22]. Apparently, all of the carboxylate polyanions mentioned above act in a similar way and the presence of base residues is only a secondary factor. Thus, there are not many possibilities to achieve template-specific inhibition with polyanionic analogs; rather their effects depend on strong coulombic interactions between the polymer and enzyme.

Specific inhibition of templated biosynthesis by electroneutral polynucleotide analogs can be achieved even with complex templates[19]. Messenger RNA coding for globin, similarly to other messenger RNA molecules, contains a polyadenylate sequence located at its 3'-terminal. This RNA molecule can be transcribed into the corresponding DNA sequences by reverse transcriptase after the addition of a suitable primer, e.g., oligothymidylate. Transcription is a unidirectional process; it starts at the 3' end of RNA, where the oligothymidylate primer is bound to polyadenylate sequences, and proceeds to the 5' end of the molecule. Poly-1-vinyluracil inhibits this reverse transcription quite effecitvely; a concentration of 14 μg/ml of polymer decreases the reaction rate to one-half. Globin's messenger RNA can, of course, be translated *in vitro* into globin. Translation also is a unidirectional process but it starts at the 5' end of RNA and no primer is necessary. The polyadenylate sequence is at the other end of messenger RNA and is not translated into protein; consequently, whatever is done with that sequence can have only indirect effects on the efficiency of the translation. Accordingly, the process was only slightly influenced by the analogs; 50% inhibition by poly-1-vinyluracil occurred only at the concentration of 1400 μg/ml.

The inhibitions described above occurred only when the analog and polynucleotide contained complementary bases. These combinations are not the only ones in which the interaction can occur, e.g., affinity methods detect some interaction between the non-complementary poly-9-vinyladenine and polyadenylate[11]. Apparently, such complexes are too unstable to affect the enzymatic reactions; nevertheless, extensive modification of the analog can increase the stability of the polymer-polynucleotide complex to the point where such a polymer can effectively inhibit the reaction. Thus, omisssion of the amino group from poly-9-vinyladenine leads to poly-9-vinylpurine and the latter polymer inhibits the reverse transcription of polyadenylate and polyuridylate[23]. The introduction of a dimethylamino group in place of the amino group of poly-9-vinyladenine abolishes all of its inhibitory effects[23]. All these effects can be correlated with the ability of polymers to form complexes with templates.

4 Polynucleotide Analogs and Cells in Culture

4.1. Cells in Culture

Interactions of synthetic polymers with the organism are formidably complex. Some simplication can be achieved by the use of cells in culture instead of the organism; interaction with a single species of cell can be studied there and the effects of other anatomical elements of tissues, e.g., collagen matrix, are eliminated. Because few chemists are familiar with this useful system, a short description of it follows. Many cells from human or animal tissues can be grown or sustained in cultures. For that purpose fresh mammalian tissue is dispersed by mechanical and enzymatic means and cells are transferred into a medium that contains a complex mixture of chemicals and about 10% animal serum. Under these conditions some cells grow, i.e., have progeny. For example, human fibroblast cells that form the basis of connective tissue in the body can be obtained from skin and grown in culture for 10–60 generations. Other cells do not divide but can be kept alive and active for several days, e.g., macrophages can be obtained from the peritoneal cavity which phagocytose and destroy bacteria in the body.

Cells in culture give us the opportunity to analyze the biological responses to polymers in relatively simple conditions. In a single experiment with cells in culture 10^4–10^6

cells can be used; 10^7–10^8 cells can be maintained routinely in the laboratory. The practical aspects have been reviewed[25]. To illustrate the size and weight of such cells we can use the data collected on human fibroblasts. Such a cell line has a volume between 1900–2600 μm^3; because its density deviates only slightly from 1, the weight of the cell in picograms (pg, 10^{-12}g) is in about the same range. One such cell contains about 10 pg of deoxyribonucleic acid, 20–45 pg of ribonucleic acid, and 500–800 pg of proteins[26]. Of course, various cells differ slightly in size and composition, e.g., monocytes from human blood have a volume of between 600–950 μm^3.

When cells in culture are treated with a polymer solution the surface of each cell is exposed to the solution. The cell is isolated from its surroundings by a cellular membrane, alternatively named cytoplasmic membrane. When cells in culture are used, all interactions between polymer and cytoplasmic membranes are uniform and can be easily measured. Beyond that point, the situation remains quite complex. The interior cell is again divided by lipid containing membranes into many compartments that prevent a uniform distribution of macromolecules in the cytoplasmic space. Thus, there are cellular spaces, again separated from the cytoplasm by membranes, that can stay completely free of polymer and consequently, free of any inhibition that the polymer may exert.

Even when the polynucleotide analog reaches the space where cellular nucleic acids are located the outcome is not predictable. The cell contains an enormous variety of nucleic acids that differ considerably in sequence and structure. Few short, simple sequences are known to occur in natural nucleic acids. As mentioned above, polyadenylate sequences are known to occur in many eukaryotic messenger ribonucleic acids, i.e., those nucleic acids that serve as templates for the synthesis of proteins. Furthermore, there are several oligothymidylate sequences in eukaryotic deoxyribonucleic acids and a cytidylate sequence in some viral nucleic acids. These segments, in spite of being part of nucleic acids, probably do not carry any immediate genetic information but serve as organizational and regulatory purposes. Thus, the presently available analogs cannot be expected to block directly any immediate genetic message; maximally they can interfere in some regulatory processes governing the replication or translation of nucleic acids.

The cytoplasmic membrane not only has the function of isolating the interior of the cell from its surroundings, but it is also the organ of cell to cell communication and responsible for collecting the nutrients for the cell. These membrane functions can easily be affected by polymers.

Specific receptors are located on the cell membrane and binding of an extracellular messenger, e.g., a hormone, to these receptors activates the formation of an intracellular messenger, a process that eventually results in the observed bioeffect. There is considerable evidence that adenosine is one of the extracellular messengers of neurosystems[27]. In the central nervous system, for example, this compound inhibits activity and behaviorally has depressant effects[28]. The adenosine receptor seems to have a very specific steric requirement and, up to now, no polymeric purine derivative has been found effective in experiments *in vivo*; however, the possibility of such applications has been demonstrated using isolated heart systems[29].

In addition to its role as a messenger, adenosine has other strong effects on cells. This compound is toxic to cells in culture even at a concentration as low as 10^{-5}M, probably through its interference with pyrimidine biosynthesis in cells[30]. This toxicity is manifested only in a specifically designed system since many sera, e.g., calf or human, that are used to grow cells contain adenosine deaminase, an enzyme that immediately detoxifies the adenosine. No polymeric adenosine derivative has been studied in this system but the usefulness of such investigations is apparent.

Similarly to mammalian cells, prokaryotic systems are known to be sensitive and to transport adenosine derivatives. This was shown using purine-deficient *E. coli* strains[31] that need adenosine added to the media for growth. This system may eventually become useful for the evaluation of adenosine polymers.

4.2 Interaction of Polynucleotide Analogs with Cells in Culture

The presently known polynucleotide analogs cannot penetrate the cytoplasmic membrane and probably are unable to interact with the receptors and transport systems located there; consequently, only nonspecific interactions take place upon exposure of cells to polynucleotide analogs.

The cytoplasmic membrane is a complex organ with an overall electronegative charge due to the anionic character of the polysaccharides and proteins located there. The main element barring the polymer from penetrating the cell interior is the lipid bilayer into which the proteins and polysaccharide components of the plasma membrane are inserted. As mentioned in the introduction, synthetic compounds can penetrate into the cytoplasm of cells by phase-phase transfer. From the external aqueous media synthetic compounds are extracted into the nonpolar phase of the lipid bilayer and from there they again equilibrate with the aqueous interior of the cell. For this mode of penetration to be efficient a compound must distribute both into the aqueous and nonpolar phases. Polynucleotide analogs, e.g., poly-1-vinyluracil and poly-9-vinyladenine, like other polymers, distribute very unevenly. These polymers dissolve well in water but cannot be extracted into the nonpolar media; thus, this mode of entry into cells is not availbable to them. In addition to this inability to penetrate through the membrane, the interaction of these polymers with the membrane is also rather weak. Electroneutral polymers do not interact strongly with polysaccharides, proteins, or lipids of the membrane and thus, poly-1-vinyluracil and poly-9-vinyladenine are only weakly adsorbed to the surface of cells in culture. Using highly radioactive polymers in a serum-free media it was estimated that about 10 and 1 pg could be adsorbed per one fibroblast cell, repsectively, and binding was about saturated in 30–60 min at room temperature[32]. In their interactions with polyvinyl analogs both murine and human cells of the fibroblast type show close simi-larity[32].

The surface of cells is constantly renewed; the new plasma membrane is synthesized in the cell and translocated to the surface while parts of the existing membrane are internalized in the endocytotic process, illustrated in Fig. 4[33]. During the endocytotic process the polymer that was adsorbed on the cell surface and a small portion of the polymer in the surrounding medium enter into cells; these polymers are enclosed in endocytotic vesicles. The internalized polymers stay firmly associated with the cells. When the internalized poly-9-vinyladenine was followed for three cell divisions, the amount correlated with the number of viable cells within experimental error. Thus,no degradation or excretion of polynucleotide analogs from cells occurred. The cells containing radioactive polymer were also fractionated and distribution of radioactivity in all these fractions was measured. Results confirm that a substantial amount of the polymer was in an enveloped form , probably in endocytotic vesicles and lysosomes[32].

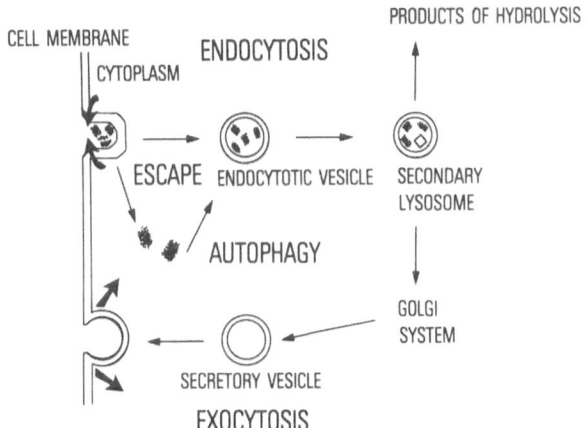

Fig. 4. Uptake and elimination of macromolecules from surrounding medium by cells. *Heavy arrows* indicate closing/opening movements of membrane leading to filling/emptying of vesicles

Polyvinyl analogs influence cells grown in culture in an interesting way. At a concentration of 0.15 mg/ml poly-1-vinyluracil is toxic to cells that are actively growing and dividing but nontoxic to stationary cultures, i.e., to cells that are not actively dividing. The toxicity of poly-9-vinyladenine is very low and cells can be grown at 1.5–2.0 mg/ml concentrations of polymer[20].

4.3 Effects on Viral Replication in Cells in Culture

Polynucleotide analogs influence replication of some viruses in cells. Poly-9-vinyladenine effectively inhibits replication of murine leukemia virus in mouse cells[20]. This virus has an interesting life cycle. The virions of murine leukemia virus contain RNA as a genetic material. After infection of cells by virus, this material is copied into DNA, a process that, as mentioned above, is called reverse transcription. Then this DNA is incorporated into the nuclear DNA of host cells. From here the genetic information of the virus is again transcribed into RNA, translated into proteins, and new viral particles are formed. Certain mouse cells already carry viral DNA in their genome, in a dormant form, but can be induced by a number of chemicals to produce viral progeny. Poly-9-vinyladenine did not inhibit his induction[34]; consequently, it can be assumed that the polymer probably inhibits some early event in viral replication, probably the reverse transcription step, which would be in agreement with the results of experiments on this enzyme[21]. The inhibition of viral growth in cells by poly-9-vinyladenine is virusspecific because replication of vesicular stomatitis virus in the same cells was not inhibited[20]. A similar compound, poly-9-vinylpurine, also inhibits replication of murine leukemia virus[35]. With this polymer it was established that the adsorption and binding of virus particles to cells was not inhibited in the presence of the polymer. Thus, viral replication must be inhibited in some subsequent step in the viral replicative cycle. Still another purine containing polymer, poly-9-vinyl-6-dimethylaminopurine, was investigated. This compound had no effect on the replication of murine leukemia virus. It is interesting to note that this polymer also does not form complexes with polynucleotides tested and does not inhibit reverse transcriptase; thus, the ability of the polymer to inhibit the viral enzyme coincides with its potency inhibiting viral replication, a coincidence that suggests the virus replication is indeed inhibited in this step[35].

The observed selective toxicities (i.e., to virus compared to cells), when considered together with the results obtained on the effects of polyvinyl analogs on individual polymerases, present an intriguing picture. Polyvinyl analogs are template-specific rather than polymerase-specific inhibitors; all cellular and viral polymerases tested were inhibited to about the same extent[17]. Thus, polyvinyl analogs in cell-free systems do not distinguish between cellular and viral enzymes; nevertheless, in cellular systems recognition occurs. This is probably due to the uneven distribution of the polymer in the cell. Viral replication occurs in the cytoplasm of cells whereas cellular replicative processes occur in the cell nucleus. Nuclear materials are isolated from the cytoplasm by an additional membrane that has limited permeability; it has pores estimated to be 45 Å in size[36]. Possibly the cytoplasm concentration of the polyvinyl analog is higher than the nuclear concentration and this forms the basis for selective toxicity. It is important to note that we do not know how the polynucleotide analog gets into the cytoplasm to inhibit viral replication there. As already mentioned, polymers are unable to penetrate the lipid bilayer of the membrane in order to penetrate into the cytoplasm. Endocytosis leads to the uptake of the polymer into the cytoplasm; however, the polymer is enclosed in the membrane in enveloped endocytotic vesicles and thus, the polymer is again fully isolated from cytoplasmic processes. Perhaps during the closure of the endocytotic vesibles, when continuity of the membrane must be disrupted, a small fraction of the polymer escapes into the cytoplasm (Fig. 4). Another possible mechanism of the entry of polymers into the cytoplasm may be through localized damage in the membrane.

Entry of the macromolecule into the cytoplasm is a process of very low efficiency, however it is known to occur with various materials, e.g., viral infection can be achieved by exposing cells to some viral nucleic acid in the presence of a basic polymer or at high concentrations of calcium ions in the medium[37]. In a latter part of this review we will present evidence to the point that foreign macromolecules free in the cytoplasm slowly undergo the autophagy process in which they are converted into an enveloped form.

4.4 Interferon Induction

Replication of various viruses in cells can also be inhibited by exposure of cells to interferons. Interferons are glycoproteins that are formed by the cells themselves when they are exposed to specific inducers. Peripheral blood leucocytes form α-interferon and γ-interferon, whereas fibroblastic cells form β-interferon. The most effective synthetic inducers of interferon are found among nucleic acids and polynucleotides, e.g., a double-stranded helical complex of polycytidylate with polyinosinate. Electroneutral polyvinyl analogs of nucleic acids are incapable of inducing interferon in human fibroblast cells in culture[38]. On the other hand, the complex of polyvinyl-1-cytosine with the polynucleotide, polyinosinate, is a very effective inducer of interferon in fibroblast human cells[9]. A reason for this effectiveness may be the high uptake of this complex by cells[38]. Induction of interferon in lymphocytes in culture by vinyl analogs of polynucleotides was investigated with similar results[39]. Also, work on mouse cells (L 929) in culture yielded similar conclusions[40, 41]. It is interesting to note that the same complex also considerably increases cellular uptake of foreign proteins from the surrounding medium[42].

5 Polynucleotide Analogs and the Organism

In studies on the effects of polymers on cells in culture only a few human and mouse cell types were used. Organisms contain a very large variety of cells that differ in their interaction with soluble macromolecules. Some cells, like macrophages and the cells of the reticuloendothelial system, have very active endocytosis, whereas other cells, e.g., fibroblast, endocytose only to a minimal extent. Of course, this complicates the interpretation of the results of *in vivo* studies. Furthermore, the organism has several supracellular barriers to polymer penetration, e.g., the barrier between the content of the alimentary tract and blood and the barrier between blood and brain. Such barriers are practically impenetrable to polymers.

The overall response of cells in culture to exposure to a polymer and the overall response of an organism to an injected polymer are quite different. Cells in culture bind only very weakly the electroneutral polymers. Thus, large amounts of highly radioactive polymers have to be used in binding studies and only a very small fraction of those stay associated with the cultured cells; the majority of unwanted foreign material remains in the solution. If the polymer is introduced into the organism proper, e.g., by injection, it is no longer a material that may be left intact, i.e., in the cell media. The injected polymer represents a circulating internal pollutant that the organism must take care of. In the organism all of the injected polymer must be effectively and substantially cleared from circulation. Thus, all the injected radioactive polymer is processed in one way or another. Organisms are equipped to deal with such internal pollutants that in natural life come from injuries, infection, and cell death and handle water soluble polymers in the same way.

The distribution of synthetic polymers in the organism has been studied only on a rather elementary level. In studies done up to now various organs have been treated as homogenous entities, while in reality, they are composed from different cells that vary considerably in their interaction with polymers.

Injection is the only effective route for introducing polymers into the organism. Polymers do not penetrate the skin and the uptake of foreign macromolecules from food is negligible.

The distribution of injected polyvinyl analogs of polynucleotides was studied in mice[32]. After injection into the peritoneum, the polymers entered the bloodstream in a matter of hours, were subsequently distributed, and partly excreted from the organism. About one third of the nondegraded polymers were excreted in the urine during the first two days. In urine the polymer was found to be of high molecular weight, i.e., it was nondialyzable. Simultaneously, with excretion processes polymers were cleared from the bloodstream by endocytosing cells located in organs; thus, in the circulation the level of polymer was found to decrease quite rapidly[32]. After the first two days the distribution of polyvinal analogs in the tissues changed only moderately; however, even after several weeks, small changes were observed. Polymers concentrate in organs that have endocytosing cells; considerable amounts of polymers were found in liver, spleen, thymus, and bone marrow. Some tissues, e.g., kidneys and lungs, were found to clear themselves of the polymer, whereas it accumulated in the aforementioned tissues. In the brain the polymer concentration was always found very low and the amounts detected there were close to observation limits (Probably derived from the blood present in excised brain).

Apparently, mammals do not possess any enzymes capable of the degradation of vinyl polymers. Two weeks after the injection, polyvinyl analogs from liver and spleen were compared with the original sample; distribution of molecular weights was found identical whithin the limits of experimental error[32].

Poly-9-vinyladenine was found nontoxic to mice when administered by intraperitoneal injection at all obtainable doses. Effects of this compound on the immunosystem, on viral leukemia, on chemically induced leukemia, and on the infection by lytic virus were investigated in detail[43].

Poly-9-vinyladenine was found to have no influence on either humoral or cell-mediated immunity. This finding contrasts with the strong influence of polynucleotides or anionic polymers on the immunosystem; these potentiate humoral responses and also influence macrophages[37].

Poly-9-vinyladenine inhibits replication of leukemia viruses in cells in culture; this polymer also accumulates and is stored in the spleen, the location where leukemia viruses replicate. Consequently, rather strong and protracted antiviral effects could have been expected. Different protocols of administration of the polymer and virus were tested and results were more intriguing than impressive. The polymer had to be administered daily to suppress viral replication; large doses administered once only were ineffective. Thus, there were antiviral effects but definitely not of the protracted character[43]. This interesting finding was explained in the following way. The polymer is taken by the spleen cells by endocytosis, the majority of it in enveloped form; thus, it is isolated by a membrane from the cytoplasm where viral replication occurs. Apparently, during endocytosis some fraction of the polymer that is taken up by cells enters the cytoplasm in a free, non-enveloped form and can there inhibit replication of the virus (Fig. 4). This free polymer is eventually converted into membrane-enveloped form; thus, at the end, all of the polymer is isolated from the site of viral replication by the membrane. There is no direct evidence of the occurrence of intracellular encapsulation of synthetic polymers; however, the phenomenon in question, called autophagy, is well known to occur with natural compounds. During cell life cytoplasmic proteins are constantly renewed, the old being enveloped into membranes, then transferred into lysosomes, and there degraded to amino acids. When synthetic polymer enters the cytoplasm, it probably follows the same pathway, i.e., it enters lysosomes and, because it is non-degradable, it is simply stored there.

Removal of the polymer from the site of its antiviral activity to an insulated site is an interesting and unusual process, leading to a loss of beneficial drug effects. Small molecular weight drugs are excreted of metabolically transformed into inactive compounds; a polymer that cannot be excreted or degraded is converted into inactive form by autophagy.

In mice poly-9-vinyladenine was found inactive against transplanted leukemia of chemical origin. Also, no inhibition of the replication of Semliki forest virus in mice was observed[43]. This virus is lytic and, similarly to leukemia viruses, it contains RNA as a genetic material. However, in difference to leukemia viruses, the life cycle of this virus does not include reverse transcription of RNA into the corresponding DNA. These results suggest that the effects of poly-9-vinyladenine on the replication of leukemia viruses in animal may also be due to direct inhibition of reverse transcription, which is a vital and unique step in the life cycle of this particular virus group.

6 Conclusion

Elegant designs in the field of bioactive polymers often result in the preparation of off-white solids (i.e., in German terminology "Schwarze Schmiere") that cannot be dissolved or biologically tested. Fortunately, synthetic analogs of polynucleotides are colorless and well soluble, and have been extensively investigated (especially the vinyl analogs). Vinyl analogs interact in a Watson-Crick manner with natural polynucleotides and can serve as templates in chemical condensation of nucleotides. Vinyl analogs cannot serve as templates in the biosynthesis of nucleic acids or proteins and, through the formation of complexes with natural polynucleotides, may prevent the latter from serving as templates in biosynthesis. Thus, vinyl analogs are true template-specific inhibitors in nucleic acid directed biosynthesis. The effects of polynucleotide analogs on cell containing systems are not pronounced. Polynucleotide analogs are prevented, by their high molcular weight and their inability to dissolve and penetrate through lipid membranes, from getting to intracellular locations where they can be truly effective. Since even poets[1] recognize the potential òf bioactive polymers there should be a way of overcoming these problems. Perhaps structural elements can be incorporated into polynucleotide analogs that would make them both water and lipid soluble; perhaps use of oligomers rather than polymers would result in the solution of these problems.

7 References

1. Overberger, C. G., Michelotti, F. W.: J. Am. Chem. Soc. *80*, 988 (1957)
2. Hoffmann, S., Witkowski, W.: Wirkungsmechanismen von Herbiciden und Synthetischen Wachstumregulatoren, p. 291, RGW-Symposium Halle 1972, VEB Gustav Fischer 1975
3. Hoffmann, S.: Z. Chem. *19*, 241 (1979)
4. Kaye, H.: J. Am. Chem. Soc. *29*, 5777 (1970)
5. Pitha, J., Pitha, P. M., Ts'o, P. O. P.: Biochim. Biophys. Acta *204*, 39 (1970)
6. Pitha, P. M., Pitha, J.: Biopolymers *9*, 965 (1970)
7. Pitha, J., Pitha, P. M., Stuart, E.: Biochemistry *10*, 4595 (1971)
8. Pitha, P. M., Michelson, A. M.: Biochim. Biophys. Acta *204*, 381 (1970)
9. Pitha, J., Pitha, P. M.: Science *172*, 1146 (1971)
10. Reynolds, F. et al.: Biochemistry *11*, 3261 (1972)
11. Pitha, J.: Anal. Biochem. *65*, 422 (1975)

---

1 G. Kunert: Aqua destillata gebiert nichts: Reinheit ist unfruchtbar. Studiere den Regen: jedes Tröpfchen ist wahr. In translation: Bioeffects unachievable with chemical individua will be obtained with polymeric mixtures

12. Chou, H. J., Froehlich, J. P., Pitha, J.: Nucl. Acids Res. *5*, 691 (1978)
13. Boguslawski, S., Olson, P. E., Mertes, M. P.: Biochemistry *15*, 3536 (1976)
14. Olson, P. E. et al.: Biochemistry *14*, 4892 (1975)
15. Maggiora, L., Boguslawski, S., Mertes, M. P.: J. Med. Chem. *20*, 1283 (1977)
16. Cowling, G. J., Jones, A. S., Walker, R. T.: Biochim. Biophys. Acta *254*, 452 (1971)
17. Pitha, J., Wilson, S. H.: Nucl. Acids Res. *3*, 825 (1976)
18. Hoffmann, S. et al.: Z. Chem. *16*, 322 (1976)
19. Reynolds, F. H. et al.: Mol. Pharmacol. *11*, 708 (1975)
20. Pitha, P. M. et al.: Proc. Natl. Acad. Sci. USA *70*, 1204 (1973)
21. Pitha, J.: Cancer Res. *36*, 1273 (1976)
22. Hoffmann, S. et al.: Z. Chem. *16*, 402 (1976)
23. Pitha, P. M., Pitha, J.: Pharmac. Ther. A. *2*, 247 (1978)
24. Papas, T. S., Pry, T. W., Chrigos, M. A.: Proc. Natl. Acad. Sci. USA *71*, 367 (1974)
25. Jacoby, W. B., Pastan, I. H.: Meth. Enzymol. *58* (1979)
26. Schneider, E. L., Mitsui, Y.: Proc. Natl. Acad. Sci. USA *73*, 3584 (1976)
27. Baer, H. P., Drummond, G. I. (eds.): Physiological and Regulatory Functions of Adenosine and Adenine Nucleotides, Raven Press, New York 1979
28. Daly, J. W., Bruns, R. F., Snyder, S. H.: Life Sci. *28*, 2083 (1981)
29. Olsson, R. A., Davis, C. C., Khoun, E. M.: Life Sci. *21*, 1343 (1977)
30. Ishii, K., Green, H.: J. Cell. Sci. *13*, 429 (1973)
31. Thomas, G. A., Varney, N. F., Burton, K.: Biochem. J. *120*, 117 (1970)
32. Noronha-Blob, L. et al.: J. Med. Chem. *20*, 356 (1977)
33. Silverstein, S. C., Steinman, R. M., Cohn, Z. A.: Ann. Rev.Biochem. *46*, 669 (1977)
34. Pitha, P. M., Pitha, J., Rowe, W. P.: Virology *63*, 568 (1975)
35. Pitha, J., Wilson, S. H., Pitha, P. M.: Biochem. Biophys. Res. Commun. *81*, 217 (1978)
36. Paine, P. L., Moore, L. C., Horowitz, S. B.: Nature *254*, 109 (1975)
37. Pitha, J.: Nucleic acids and sulfate and phosphate polyanions, in: Anionic Polymeric Drugs (eds.) Donaruma, L. G., Ottenbrite, R. M., Vogl, O., p. 277, New York, John Wiley & Sons, Inc. 1980
38. Noronha-Blob, L., Pitha, J.: Biochim. Biophys. Acta *519*, 285 (1978)
39. Waschke, K. et al.: Acta Biol. Med. Germ. *38*, 739 (1979)
40. Waschke, K. et al.: Arch. Immunol. Ther. Exp. *25*, 627 (1977)
41. Hoffmann, S. et al.: Z. Chem. *17*, 61 (1977)
42. Ryser, H. J. P., Termin, T. E., Barnes, P. R.: J. Cell Physiol. *87*, 221 (1976)
43. Vengris, V. E. et al.: Mol. Pharmacol. *14*, 271 (1978)
44. Takemoto, K., Inaki, Y.: Adv. Polym. Sci. *41*, 1 (1981)

Received July 28, 1982
C. G. Overberger (Editor)

Photochromic Phenomena in the Solid Phase

G. Smets

Laboratory of Macromolecular Chemistry, K. Universiteit Leuven, Belgium

Photochromic phenomena occurring in the solid phase are discussed first from the point of view of the influence of a polymer medium on the photochromic behavior and reaction reversibility. Inversely, in the second part, the influence of conformational changes of chain segments induced by chromophore isomerization is discussed with respect to the physical behavior of the polymer matrix, and especially concerning photomechanical effects.

Advances in Polymer Science 50
© Springer-Verlag Berlin Heidelberg 1983

1 Introduction

This article reviews the behavior of photochromic macromolecules in the *solid state* and the problem has to be considered from two different points of view:

To what extent are isomerization and the resulting conformational changes of chromophores hindered by their incorporation in a solid matrix? This first question concerns the influence of the physical properties of the polymer medium on isomerization reactions.

Inversely, can we expect any change of physical property of a photochromic polymer in bulk under irradiation by light absorption?

Both aspects will be examined successively.

It will be shown in a forthcoming review[1] that a reversible change of conformation of linear macromolecules *in solution* may result in an appreciable change of solution properties, and particularly of their viscosity. Such conformational changes and the resulting effects can be displayed by chemical means, and were actually described several years ago by Kuhn, Katchalsky and coworkers[2-6] and more recently by Osada and Saito[7]. They are responsible for the mechanochemical behavior of polymer systems in the solid state, i.e. the conversion of chemical into mechanical energy[8-15].

It was only recently that these phenomena were detected photochemically using photochromic molecules whose chromophores undergo a reversible photoisomerization reaction. The photochemical reactions involved in these processes are mainly the trans-cis isomerization of aromatic azo compounds[16-20] and stilbene derivatives[21] as well as the ring opening/closure reaction of spirobenzopyran derivatives[22-24].

2 Photochromism in Bulk Polymers

Most elaborated studies were based on the reversible isomerization of aromatic azo compounds (Eq. (1)) and the ring opening/closure of spirobenzopyrans (Eq. (2)). While in the first case, isomerization can proceed without important steric requirement, in the second case photochromism involves cleavage of the C–O-pyran bond followed by rotation of one part of the molecule so as to approach coplanarity; the change of conformation is therefore more pronounced.

$$\text{Ar–N} \overset{}{\underset{}{\text{N–Ar}}} \rightleftharpoons \begin{array}{c} \text{Ar–N} \\ \| \\ \text{Ar–N} \end{array} \tag{1}$$

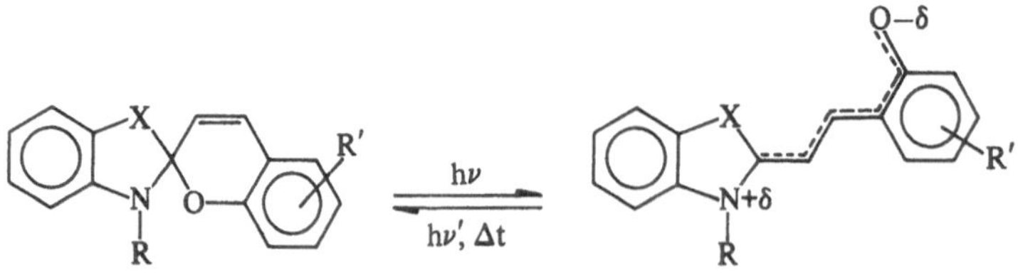

$$\tag{2}$$

where X is S, O, CMe$_2$

2.1 Rates of Decoloration

2.1.1 Experimental Results

As a general statement, isomerizations occur much slower (around 100times and more) in the film than in solution. It was already observed for azo compounds by Kamogawa et al.[25] in the case of copolymers of 4-vinyl-4'-dimethylaminoazobenzene (I) with styrene and of 4-acryloylaminomethylaminoazobenzene (II) with styrene, butyl acrylate and methyl methacrylate.

$$4\text{-}CH_2\!=\!CH\text{-}C_6H_4\text{-}N\!=\!N\text{-}C_6H_4\text{-}NMe_2\text{-}4'$$
$$I$$

$$4\text{-}(CH_2\!=\!CH\text{-}CO\text{-}NH\text{-}CH_2\text{-}NH)\text{-}C_6H_4\text{-}N\!=\!N\text{-}C_6H_5$$
$$II$$

By measuring recovery half-life periods, they found that the reaction was slowest with comonomers which increase the rigidity of the chain, e.g. styrene or methyl methacrylate compared to methyl acrylate. In a similar way, the rate of fading of a merocyanine, obtained by ring opening of a spirobenzopyran, is 100–400times smaller in a poly(methyl methacrylate) film than in an homologous solvent[26, 27]. Moreover, due to the strong negative solvatochromism of these compounds, their rate of decoloration depends on the nature of the matrix. Thus, for the 1-(β-isobutyryloxyethyl)-6'-nitro-DIPS[1](III) the rate of fading at room temperature is 8 to 10times higher in polystyrene than in poly(methyl-methacrylate) (2 wt.%); moreover, it decreases with increasing photochrome concentration in the film (for further details see Table 1, page 25).

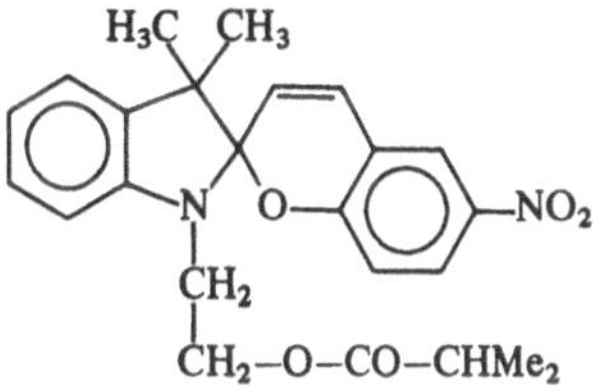

$$III$$

The dependence of the rate of decoloration on temperature, especially around the glass temperature, will be considered in Sect. 2.2[28, 29].

It has been observed for spirobenzopyrans as well as for azo compounds that, when dissolved in bulk polymers or linked to the latter as side groups, the plot of the logarithm of the optical density versus time shows strong deviations from first-order kinetics. Thus, the decoloration of a film of a copolymer of 1-(β-methacryloyloxyethyl)-6'-nitro-DIPS with methyl methacrylate after irradiation is represented in Fig. 1[29].

Such rate of decoloration was initially interpreted as being the sum of two or three exponential rate processes, each characterized by its own rate constant ($k_1 > k_2 > k_3$), which should correspond to a different merocyanine isomer (Fig. 2).

---

1 DIPS designates 3,3-dimethyl-spiro-2 H-1-benzopyran-2,2'-indoline

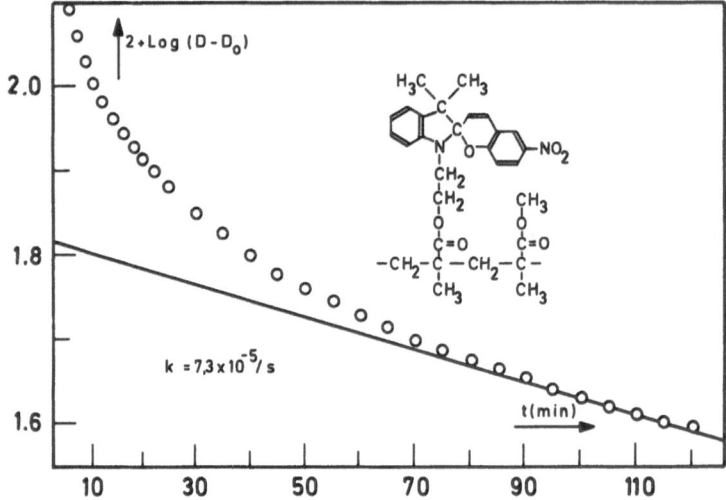

Fig. 1. Decoloration of a photochromic copolymer (27 °C) Determination of the rate constant k_3 (slowest step) λ_{max}: 585 nm

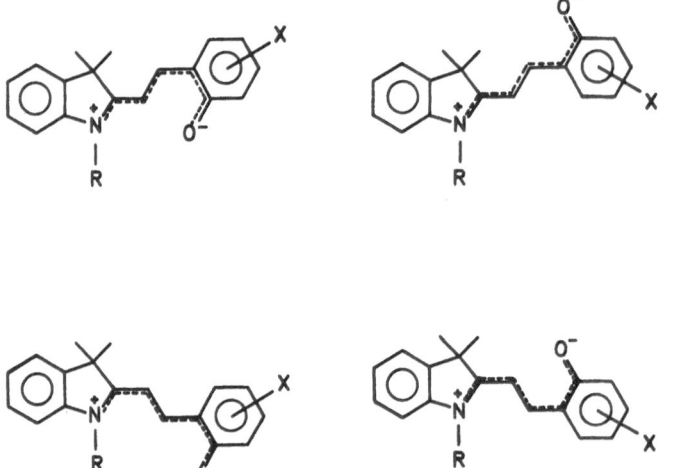

Fig. 2. Stereoisomers of merocyanine derived from X-substituted 1-alkyl-3,3-dimethylindolino-spirobenzopyran

The existence of four merocyanine isomers had indeed been demonstrated previously[30, 31]. The decoloration rate equation can be written

$$[A]_t = [X_1]e^{-k_1t} + [X_2]e^{-k_2t} + [X_3]e^{-k_3t} \tag{3}$$

in which the [X]'s represent the contributions of each isomer to the optical density $[A]_0$ at time zero.

More recently, a similar photochromic behavior was observed by Eisenbach[32, 33] for the cis-trans isomerization of azobenzene residues attached as side groups to copolymers when examined below their glass transition temperature. These copolymers were obtained by copolymerization of the monomers 4-methacryloylaminoazobenzene (IV) and 4-(3-methacryloyloxybutylcarbonylamino)azobenzene (V) with various alkyl acrylates and methacrylates.

$$C_6H_5-N=N-C_6H_4-(NH-CO-\overset{\overset{\displaystyle CH_3}{|}}{C}=CH_2)-4$$
$$IV$$

$$C_6H_5-N=N-C_6H_4-(NH-CO-(CH_2)_3-O-CO-\overset{\overset{\displaystyle CH_3}{|}}{C}=CH_2)-4$$
$$V$$

In solution and in the rubbery state, the thermal cis-trans recovery follows first-order kinetics; in the glassy state, however, some azo groups react anomalously fast, others isomerize much slower, nearly as in solution. These data confirmed earlier observations of Paik and Morawetz in the case of methyl methacrylate-styrene copolymers containing azobenzene side groups[34].

Here, also the decoloration curves can be resolved by two simultaneous first – order reactions, as above for spirobenzopyrans.

2.1.2 Interpretation

For the decoloration of merocyanines, the first interpretation assumed the existence of different isomers which each should fade, independently following first-order kinetics. It was based on the presence in some cases of two absorption band maxima in the visible spectrum after irradiation (e.g. for photochromic polytyrosines[28]) and on the changes of the relative intensities of these absorption bands during decoloration[35, 36] which were sometimes observed even in solution. Noteworthy in this respect is the behavior of 1,3,3-trimethylindolinospironaphthopyran dissolved in bulk polystyrene[37]. On irradiation, not only two absorption maxima are formed (535 and 565 nm) but, on continuing irradiation, the relative absorption at the higher wavelength maximum increases while the decoloration kinetics change appreciably. In agreement with the interpretation of Fischer et al.[38] it is assumed that in a rigid matrix the rate of fading is slower because of restrictions of rotation which prevent the merocyanine from reaching the necessary conformation for ring closure, and that these restrictions may be different for the isomeric merocyanines. The following reaction mechanism was proposed

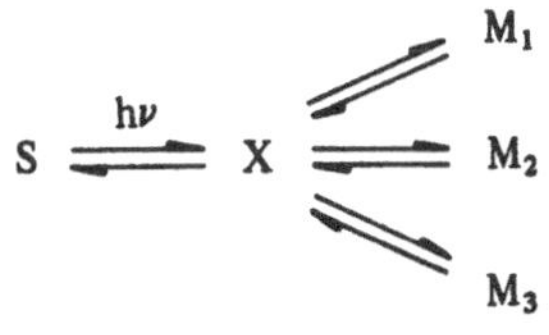

$$S \underset{}{\overset{h\nu}{\rightleftharpoons}} X \underset{\begin{array}{c}\nearrow M_1 \\ \rightleftharpoons M_2 \\ \searrow M_3\end{array}}{} \qquad (4)$$

in which S denotes the closed spironaphthopyran, M_1, M_2, M_3 are isomeric colored forms and X the reaction intermediate. The transformation of X to S is assumed to be very rapid in such a way that the rate-determining step of decoloration is the transformation of the M's into X; the thermal interconvertibility of the isomers should be hindered by the high internal viscosity of the solid polymer. Very likely, these observations indicate the existence, at least in some cases, of different isomers which fade at different rates. The interpretation suffers, however, from severe objections, especially in systems where only one active isomer has to be accepted (e.g. azo derivatives) and where similar deviations from first-order kinetics are also observed below the glass transition temperature T_g[32]. Moreover, on considering the structures of different stereoisomers it seems doubtful that the interconversion of the different isomers (i.e. trans → trans isomerization) should be prevented in the solid polymer while their trans → cis isomerization followed by ring closure (Eq. (2)) is possible and is actually observed. Yet on the basis of electrochromism experiments in which the equilibrium between the merocyanine isomers was disturbed by the application of a strong electric field, Kryszewski and Nadolski[39, 40] observed that the rate constants of trans → trans isomerization of 6'-nitro-DIPS and of 6'nitro-8'methoxy-DIPS dissolved in poly(methyl methacrylate) and poly(butyl methacrylate) are much higher than those of ring closure. The ensemble of these experimental data and considerations leads to the conclusion that the anomalous kinetic behavior of the photochromes below T_g is mainly due to a non-homogeneous distribution of the free volume in the polymer matrix and thus to the physical state of the glassy polymer[32, 34, 41, 42] and that the chemistry of the reaction is usually not rate-determining and consequently of secondary importance to the photochromic behavior. This conclusion is strongly supported by the experimental results of Smets et al.[43] concerning the decoloration behavior of photochromes of different molecular size in a polymer matrix, and the influence of film stretching on these phenomena[44].

They compared 1-benzyl-6'-nitro DIPS (VI) with 1,1-xylylene-bis-DIPS derivatives, namely its bispropionate (VII), oligo- and polyesters (VIII) in polystyrene and poly(ethy-

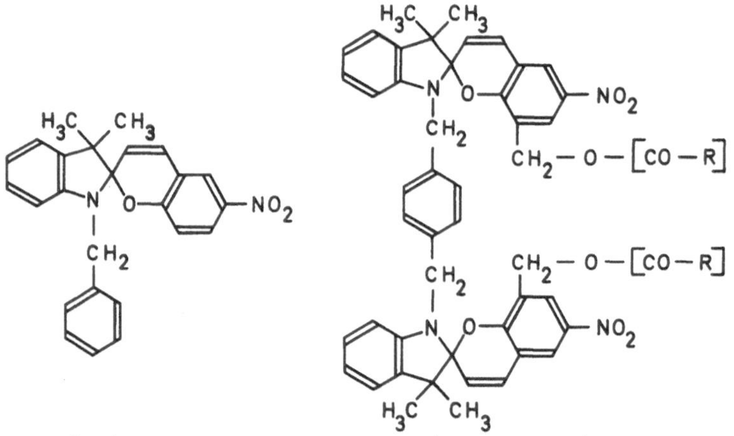

Fig. 3. 1-Benzyl-6'-nitro-DIPS (VI) and 1,1-xylylene-bis-(6'-nitro-DIPS) analoges, where R = C_2H_5: bispropionate ester (VII)

$$R{-}CO{-}is\left[\begin{array}{c} CH_3 \\ | \\ CO{-}(CH_2)_n{-}CO{-}O{-}C_6H_4{-}C{-}C_6H_4{-}O \\ | \\ CH_3 \end{array}\right]_x H \qquad \text{bis-polyester (VIII)}$$

lene glycol terephthalate/isophthalate)[50/50]. The structural analogy of these compounds should be underlined, as can be seen from following structural formula.

For the bispropionate –CO–R designates –CO–C$_2$H$_5$ (VII). For the oligo- and polyesters (VIII) –CO–R is

$$\left[-CO-(CH_2)_n-CO-O-C_6H_4-\overset{\overset{\displaystyle CH_3}{|}}{\underset{\underset{\displaystyle CH_3}{|}}{C}}-C_6H_4-O-\right]_x H$$

VIII

for oligomers n = 4 and x varies between 2 and 8; for polymers n = 5 and x ranges from 20 to 25.

The decoloration kinetics are represented in Fig. 4 where the logarithm of the optical density is plotted against time.

The bispropionate dimer decolorizes already much slower than the monomeric model, but more rapidly than the oligomers and polyesters in which the bisphotochromes are incorporated in the polymer backbone, and which behave almost identically. For the latter compounds, the optical density decreases during the first hours after irradiation, then remains almost constant for very long periods and fading becomes almost negligible; it will only disappear by heating the film above its T$_g$.

On the other hand, stretching the polymer matrix causes molecular orientation of the photochrome as shown by dichroism measurements; it will also restrict their isomerization possibilities[44]. Thus, stretching of a film of poly(bisphenol-A-pimelate) (M$_n$: 32 000)

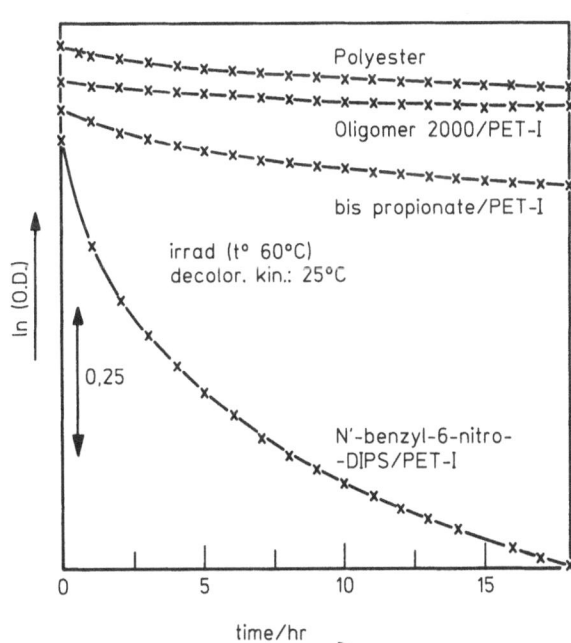

Fig. 4. Decoloration kinetics of photochromes at 25 °C in a polyterephtalate/polyisophtalate (50/50) matrix. Influence of the molecular size of the photochrome. Irradiation above T$_g$ (60 °C)

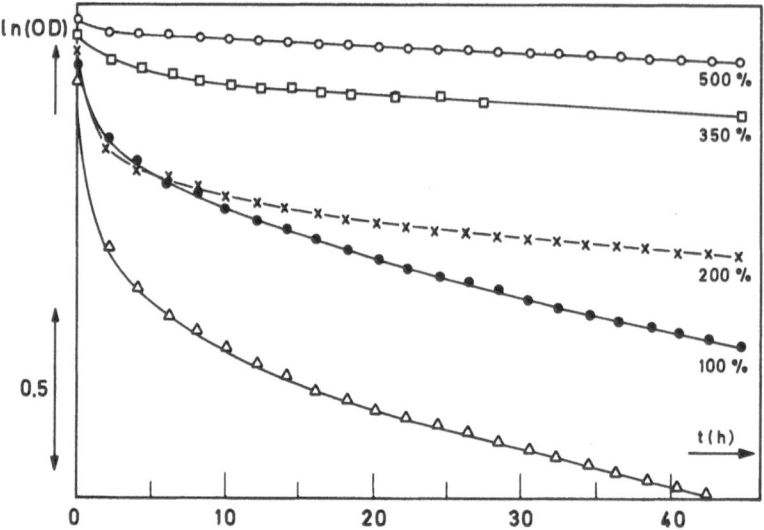

Fig. 5. Decoloration of photochrome bispropionate (VII). Influence of film stretching

containing 5wt-% bisphotochrome propionate affects considerably the first rapid decoloration phase as well as the second phase that becomes almost negligible at high elongation (Fig. 5).

Such molecular orientation effect on colour fading occurs only temporarily with the monophotochrome (VI) and vanishes indeed progressively on storage[43]. In contrast, as soon as the molecular size of the photochrome becomes sufficiently high by increasing the length of the ester groups attached in position 8' (formula VIII, x > 2) thermal fading becomes very slow and makes stretching of the matrix useless. These experiments show clearly the preponderant influence exerted by the molecular size of the photochrome on thermal fading; by choosing adequately the polymer matrix and the molar size of the photochrome thermal fading can be avoided, i.e. the thermal reversibility completely suppressed.

2.2 Influence of Temperature

2.2.1 Experimental Results

The physical interpretation of the photochromic decoloration behavior is considerably strengthened if one considers the influence of temperature on these phenomena. The activation parameters of the decoloration of 1-(β-isobutyryloxyethyl)-6'-nitro-DIPS (III) dissolved in poly(methyl methacrylate) or in polystyrene (2wt-%) are given in Table 1, in which k_2 and k_3 correspond to the apparent first-order rate constants of the second and third (slowest) stages of fading of the merocyanine[29].

Not only the activation energies differ considerably on account of the strong negative solvatochromism of these compounds, but the activation entropies have opposite signs in the two matrices. Analogous observations were already mentioned by Flannery for 1-

Table 1. Activation parameters of decoloration of 1-(β-isobutyryloxyethyl)-6'-DIPS in bulk polymers

	Polystyrene		Poly(methyl methacrylate)[a]	
	k_2	k_3	k_2	k_3
$k \times 10^4$ (s^{-1}), 27 °C)	175	17	17(8)	2(1)
E_a(kcal/mol)	16.3	17.6	22.5(24.4)	24.4(25.7)
$\Delta S^{\neq}$ (cal mol^{-1} deg^{-1})	−18.8	−14.6	+1.8	+3.9

[a] Bracketed values correspond to 10% photochrome in PMMA film

methyl-6'-nitro-DIPS in solutions, $\Delta S^{\neq}$ being positive in polar solvents and negative in non-polar ones[45]. The temperature dependence can also differ, depending on whether the photochrome is bound covalently to the polymer as a side group (through copolymerization) or if it is only dissolved in the corresponding amorphous polymer. This problem of photochromic incorporation was examined by Verborgt and Smets[29] in the case of copolymers of photochromes IX and X with methyl and isobutyl methacrylate; their decoloration behavior was compared with that of the homologous isobutyric model dissolved in the corresponding polymethacrylate (5wt-%). Both photochromes differ from each other only by their N-substituent (amide or ester linkage).

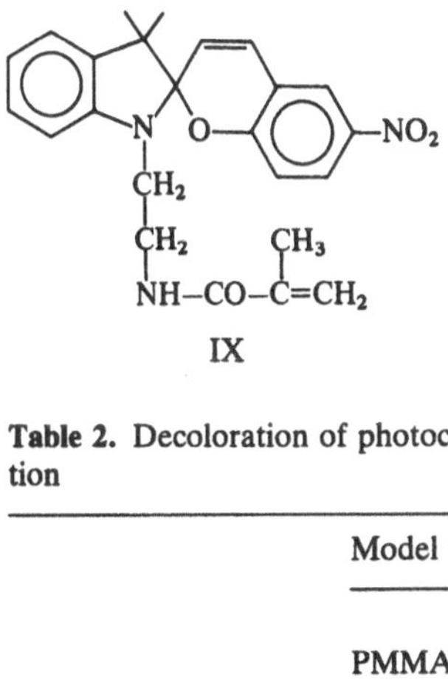

IX

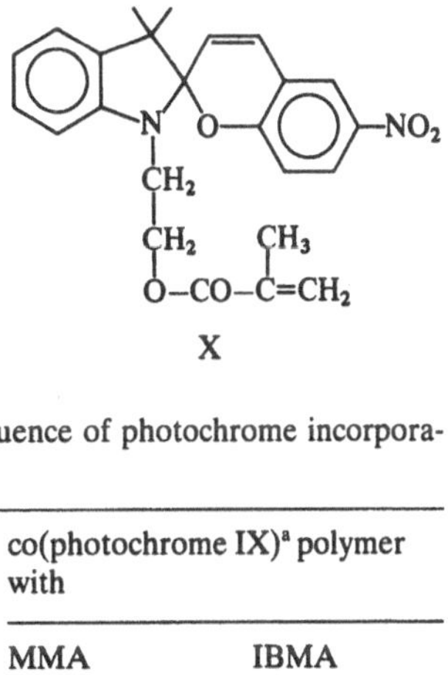

X

Table 2. Decoloration of photochrome IX in bulk polymer. Influence of photochrome incorporation

	Model photochrome (5 wt-%) in				co(photochrome IX)[a] polymer with			
	PMMA		PIBMA		MMA (4.9 wt-%)		IBMA (4.85 wt-%)	
Stage of decoloration	(2)	(3)	(2)	(3)	(2)	(3)	(2)	(3)
$k \times 10^4$ (s^{-1}),	5	0.51	13.5	1.6	10.6	0.7	34	2.7
E_a (kcal/mol)	21.5	24	25.9	32.1	16.6	19	13.7	15.9
$\Delta S^{\neq}$ (cal mol^{-1} deg^{-1})	−4.3	−0.5	11.7	21.7	−19	−16	−26	−24

[a] Photochrome content

It can be seen from Table 2 that the rate constants at 27 °C are higher for the copolymers than for the mixtures. This order is reversed at higher temperature, due to the smaller E_a-value for the copolymers. The strong negative values of the activation entropies for the copolymers should be pointed out. Worth noting is also that the kinetic parameters vary with the photochrome content in the copolymer.

Table 3. Co(photochrome IX-MMA) polymer. Influence of photochrome content

Photochrome (wt-%)	2.8		4.9		12.7	
Stage of decoloration	(2)	(3)	(2)	(3)	(2)	(3)
$k \times 10^4$ (s^{-1}), 27 °C	19	1.2	10.3	0.7	5.5	0.33
E_a (kcal mol^{-1})	15.6	18	16.6	19	18.3	20.4
ΔS^* (cal mol^{-1} deg^{-1})	−21	−18.4	−19	−16	−14.5	−13

Table 4. Decoloration of photochrome X in bulk polymer. Influence of photochrome incorporation

	Model photochrome (5 wt-%) in PMMA		Co(photochrome X-MMA) polymer (5.4 wt-%)	
Stage of decoloration	(2)	(3)	(2)	(3)
$k \times 10^4$ (s^{-1}); 27 °C	16	1.75	8.8	0.7
E_a (kcal mol^{-1})	23.4	25.1	22.2	22.9
ΔS^* (cal mol^{-1} deg^{-1})	+4.7	+5.9	−0.5	−3.2

For copolymers of photochrome X the situation is however different from the previous one: the mixtures decolorize about twice as fast as the copolymers while the activation energies differ only slightly.

Considering that the photochromes IX and X only differ in their N-substituent, the comparison of these data shows that the nature of the photochrome must also be taken into consideration, and that it would be hazardous to draw general conclusions about the influence of photochrome incorporation. In the case of azobenzene side groups (see Table 6) binding or not of the photochrome on the polymer chain does not exert a significant effect on the decoloration kinetics.

Most striking is however the dependence of the rate of bleaching in the neighborhood of the glass transition temperature of the copolymers. This dependence is illustrated in Tables 5 and 6 for methacrylic copolymers carrying spirobenzopyran and azo side groups, respectively. In these experiments the spirobenzopyran comonomer was 1-(β-methacryloylaminoethyl)-6'-nitro-DIPS (IX) and the azo derivative comonomer 4-methacryloylaminoazobenzene (IV).

The changes of the activation energy and activation entropy around the glass transition temperature listed in Tables 5 and 6 are of such importance that they can only be interpreted on the basis of physical changes of the polymeric environment, i.e. an important increase of the chain segment mobility and consequently of the free volume above T_g. The additional increase of the activation energy has to be related with the E_a of viscous flow of the polymer.

Figure 6 illustrates clearly the significance of T_g concerning the free volume available for molecular motions and ring-closure of the merocyanine side groups. Similar diagrams are obtained for azo chromophores present in several copolymers[23].

Table 5. Decoloration of photochrome VIII copolymers. Influence of the glass transition temperature T_g [29]

	Comonomer					
	Isobutyl methacrylate		Propyl methacrylate			
wt-% photochrome	2.6 (T_g 61 °C)		2.7 (T_g 53 °C)		5.3 (T_g 55 °C)	
Decoloration stage	(2)	(3)	(2)	(3)	(2)	(3)
$k \times 10^4$ (s⁻¹) 27 °C	56.5	4.5	42	3.6	31.2	2.6
64 °C	741	92				
65 °C			1280	267	942	162
69 °C			2750	560	1720	348
70 °C	1370	210				
below T_g						
E_a (kcal/mol)	12.6	15.1	15.4	17.2	17.6	19.8
ΔS^* (cal mol⁻¹ deg⁻¹)	−29	−25.5	−20.1	−19	−13.3	−19.7
above T_g						
E_a (kcal/mol)	26.4	32.3	31.5	37	28.8	36.5
ΔS^* (cal mol⁻¹ deg⁻¹)	+12.4	+25.8	+27.7	+41	+19.7	+43.6

Table 6. Decoloration of azobenzene side groups in bulk copolymers* of 4-methacrylaminoazobenzene. Influence of the glass transition temperature [32]

	Comonomer					
	Model cpd. in poly(ethylmethacrylate)		Ethyl methacrylate		Isobutyl methacrylate	
decoloration stage	(1)	(2)	(1)	(2)	(1)	(2)
$k \times 10^4$ (s⁻¹) 25 °C	0.1	4.5	0.12	4.3	0.13	5.3)
70 °C	7.2	0	4.2	20	6.7	0
below T_g						
E_a (kcal mol)	17.1	5.6	14.1	6.4	15.5	5.4
ΔS^* (cal mol⁻¹ deg⁻¹)	−27	−57	−36	−56	−31	−58
above T_g						
E_a (kcal/mol)			29.5	22.6	28.8	
ΔS^* (cal mol⁻¹ deg⁻¹)			+10	−7	+9	

* All these copolymers contain about 0.1 mol-% azobenzene chromophore, i.e. one chromophore per macromolecule. (1) and (2) represent the two stages of the decoloration, the normal slow step and the anomalously fast step, respectively

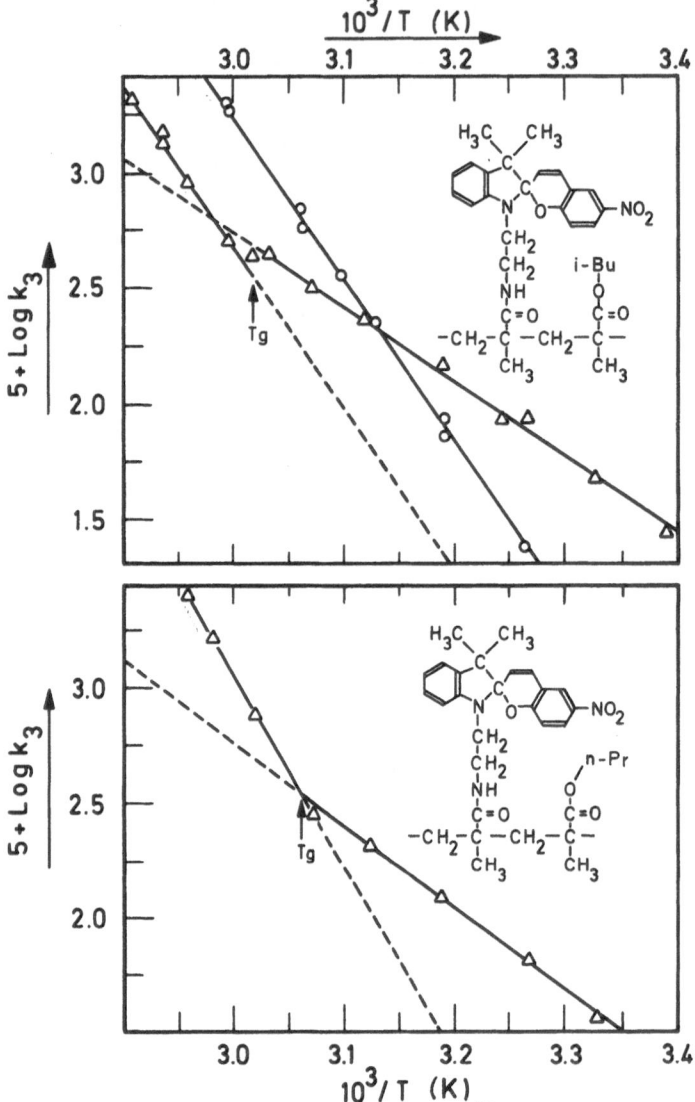

Fig. 6. Arrhenius plots of decoloration rate constants k_3 (slowest step) of copolymers of 1-(β-methacryloylaminoethyl)-6'-nitro-DIPS with isobutyl and n-propyl methacrylate ($\triangle$). Influence of the glass transition temperature.
Model photochrome 1-(β-isobutyrylaminoethyl)-6'-nitro-DIPS in poly(isobutyl methacrylate): ($\bigcirc$)

It should be expected that the influence of the chain segment mobility is most pronounced when the photochrome groups are inserted into a semirigid polymer backbone instead of being attached as mobile side groups in copolymers. Therefore, polyesters have been prepared by condensation of bis-hydroxymethyl-spirobenzopyrans with bisacid dichlorides followed by polyesterification with bisphenol-A[43, 46]. Thus, a photochromic poly(bisphenol-A-pimelate) (polyester XI) of the following formula was obtained.

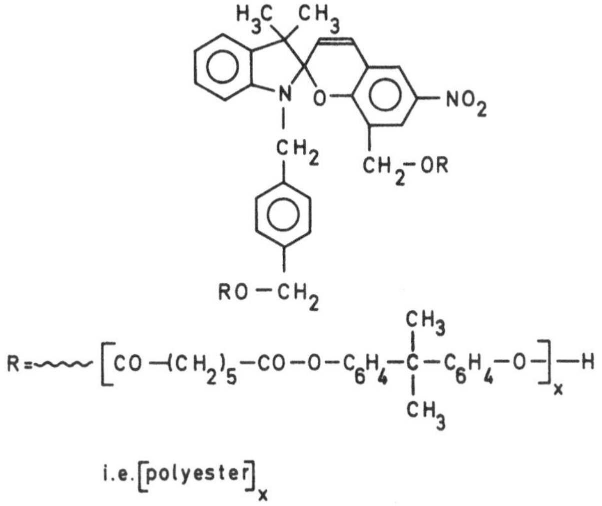

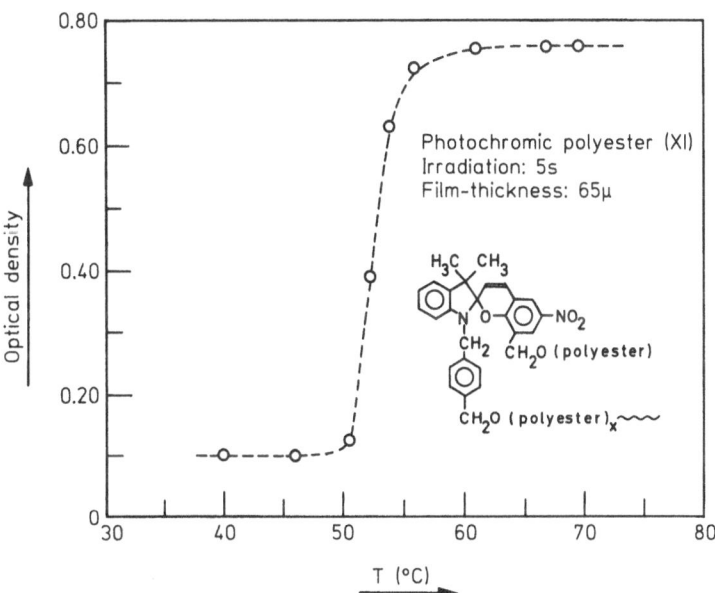

Fig. 7. Structure of photochromic poly(bisphenol-A-pimelate) (XI) (M = 13 200); T$_g$: 50–55 °C

Fig. 8. Dependence of the photoresponse of photochromic polyester XI on temperature

Its photoresponse was examined around T$_g$ (50–55 °C) where the segment mobility is assumed to be extremely sensitive to a temperature change.

Figure 8 shows the optical density of a film (65 microm. thickness) after an irradiation period of five seconds. As can be seen, a sharp increase of photoresponse occurs in the direct neighbourhood of T$_g$. On following the rate of decoloration at several temperatures an Arrhenius plot gives a sigmoid curve (Fig. 9); above 70 °C the activation energy is equal to

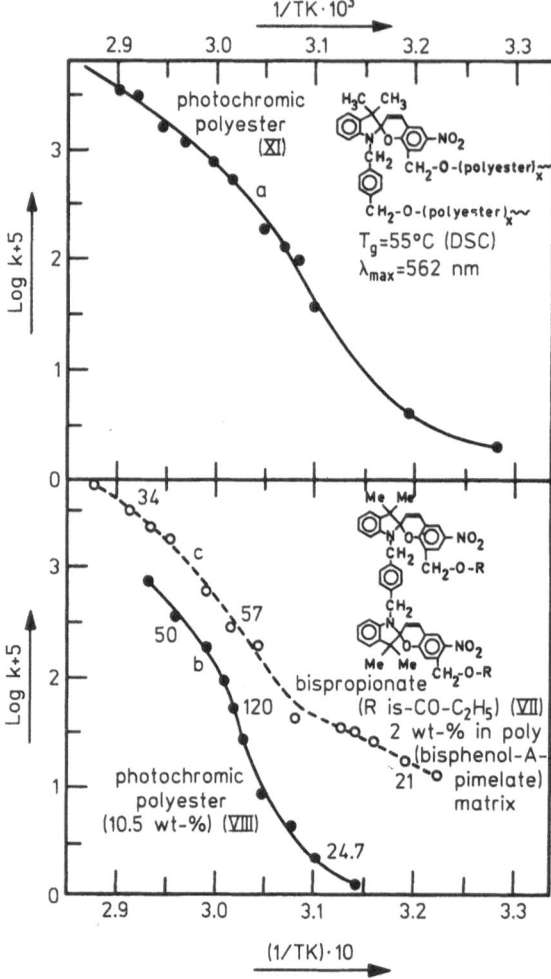

Fig. 9. Dependence of the decoloration rate constant (ln k) on 1/T for photochromic polyesters (XI-a) and (VIII-b) curve C corresponds to the model bisphotochrome propionate (VII) dissolved in the poly(bisphenol-A-pimelate) matrix (figures indicate apparent E_a-values)

29–30 kcal/mol, a value which corresponds to that of the model compound in dimethylformamide. In the neighborhood of the glass transition, the apparent activation energy is found to be equal to 76 kcal/mol.

For polyester VIII characterized by the presence of a bisphotochromic chromophore in the main chain, a still more pronounced temperature dependence should be observed. Indeed, an apparent activation energy around T_g of 120 kcal/mol emphasizes the strong influence of bulky spirobenzopyran groups in the main chain on the photochromic behavior.

2.2.2 Interpretation

The strong temperature dependence of the decoloration rate constant must undoubtedly be related to the rigidity of the polymeric matrix below the glass transition temperature which inhibits photochromic isomerization. It is only around and above T_g that the molecular mobility becomes appreciable, gradually increasing and facilitating isomerization.

Smets and Evens[46] assumed that the pre-exponential term of the Arrhenius equation of the rate constant is proportional to the jump frequency of a molecular segment from one position to another, i.e. proportional to the reciprocal of the internal viscosity of the bulk polymer at a given temperature.

It will therefore be characterized by an activation energy for the jump of a segment of a polymer chain from one equilibrium position to the next; it also depends on the local configurational arrangement of nearest neighbor segments[47].

Taking into account the well-known WLF-equation[48] relating the viscosity-temperature coefficient of a polymer to its glass temperature they obtained Eqs. (5) and (6)

$$\log(k_{T_g}/k_T) - E_a(T - T_g)/2.3\,RTT_g = -C_1(T - T_g)/(C_2 + T - T_g) \tag{5}$$

or by transformation,

$$[\log(k_T/k_{T_g}) + E_a(T - T_g)/2.3\,RTT_g]^{-1} = \frac{1}{C_1} + C_2/C_1(T - T_g) \tag{6}$$

where E_a is the *chemical* energy of activation of isomerization while k_T and k_{T_g} are the decoloration rate constants at the temperatures T and T_g respectively.

By plotting the left-hand expression of Eq. (6) against $1/(T - T_g)$ straight lines are obtained from the intercept (after extrapolation) and the slope of which the "constants" C_1 and C_2 respectively are calculated[49].

Figure 10 illustrates such a linear relationship for the bisphotochromic polyester VIII (M = 16000) for which C_1 = 9.09 and C_2 = 31.36 instead of the usual 17.44 and 51.6 values respectively[48]. Using these values of C_1 and C_2, one can recalculate the curve of ln k_T against T and T_g. This calculated curve corresponds to curve b in Fig. 9, where the points indicate the experimental log k values.

A similar but much more detailed interpretation was recently given by Eisenbach[50, 51, 52] who showed the general validity of the WLF-equation in the interpretation of the photochromism of azobenzenes and spirobenzopyrans linked to amorphous bulk polymers. Eisenbach used directly the WLF-equation[7]

$$\log a_T = -C_1(T - T_g)/(C_2 + T - T_g) \tag{7}$$

where a_T is the ratio of the relaxation time at temperatore T to that at $T_g (a_T = k_{T_g}/k_T)$. On the basis of this equation he determined graphically C_1 and C_2 for several photochromic systems. The corresponding numerical values were nearly constant for a group of polymers, e.g. for photochromes (azo and spiropyrans) dissolved in different polyacrylates and polymethacrylates and attached to the latter as side groups. They differ, however, considerably below and above the glass transition temperature of the polymer

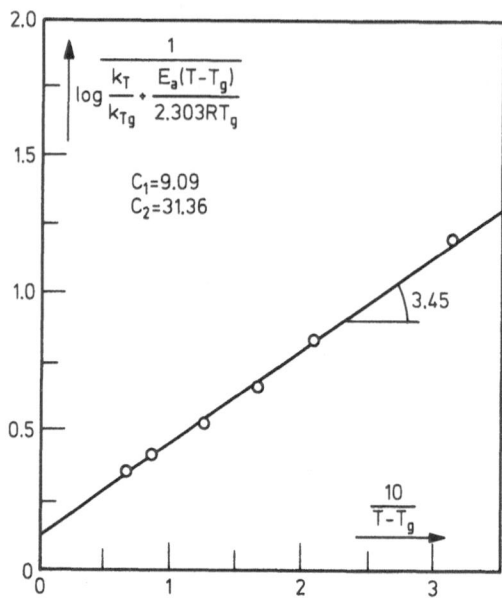

Fig. 10. Determination of the C_1 and C_2 constants using the WLF-equation for the bisphotochromic polyester VIII

system. It should be noticed that these experimental C_1 and C_2 values differ markedly from the "universal" WLF constants[48], likely because the photochromic behavior is mainly controlled by the free volume distribution around the photochrome, which is different from its usual random distribution. Moreover, irradiation causes defects in the matrix resulting from the excess of energy absorbed by the photochrome; the energy required for ring opening amounts only to 20–30 kcal, while the energy absorbed at λ : 365 nm corresponds to 78 kcal/mol. This excess of energy is transmitted to the medium and leads to conformational changes even in the glassy state.

Eisenbach also discussed the significance of the WLF-parameters with respect to the fractional free volume f_g at T_g and to the difference of thermal dilation coefficients of rubbery and glassy states.

In most cases, the fractional free volume amounts to $0.025 - 0.030$, a value which is considered as a normal one for amorphous polymers. For the photochromic polyesters inadminissibly high f_g values were however found (corrected f_g 0.13 to 0.197 instead of 0.27 to 0.45 as indicated erroneously). On the basis of the calculation of Sevens[49] f_g is 0.048 which is much more reasonable when taking into account the bulkiness of the bisphotochrome and the disturbing effects on its direct vicinity.

Similarly, Sung et al. recently reported the use of azo chromophoric labels as a molecular probe of physical aging in amorphous polymers[53]. By measuring the kinetics of trans $\rightleftharpoons$ cis photoisomerization of azo chromophores covalently bonded to amorphous polyurethanes, they defined a parameter α which corresponds to the fraction of the free volume above a critical size at a given temperature and time of aging.

Noteworthy is also the discussion of Kryszewski and Nadolski[40] on the photochromic properties of spirobenzopyran copolymers and copolyesters. They interpreted the photochromic fading on the basis of the formation of defects within the matrix resulting from the excess of energy on irradiation as mentioned above.

These primary defects would relax rapidly and be transformed into secondary defects the volume of which is related with the local volume difference between the DIPS and their open-merocyanines. Assuming a non-uniform distribution of distances between the merocyanine molecules and these secondary defects, they found indeed a time-dependent concentration of the merocyanine given by the following equation

$$\frac{(M)_t}{(M)_0} = I - K\sqrt{t}.$$

On the basis of this defect model the disappearance of 60–70% of the merocyanine molecules may be accounted for. Moreover, the constant K is a sensitive indicator of the glass transition temperature, as demonstrated on the basis of the Arrhenius plots.

From Chap. 2, it must be concluded that the photochromic isomerization reactions and their resulting conformational changes are greatly affected by the polymeric medium in which these reactions take place. It was shown that

i)   the chain segment mobility of the polymer, i.e. its glass transition temperature and chain orientation, exerts a major influence on the photochrome behavior;

ii)   photochromes can be used as a probe for the detection of local chain movements;

iii) photosensitive compounds with an appropriate structure permit to modify or even to suppress the photoresponse behavior.

3 Photomechanical Phenomena

In this chapter we will consider to which extent conformational changes of the chromophores and connected chain segments of a photochromic polymer may induce changes of dimension of bulk polymers and thus generate reversible photomechanical effects.

3.1 Cross-Linked Polymers and Swollen Gels

From the organic point of view, reversible contraction/dilation phenomena should be observed in photochromic networks below their glass transition temperatures, i.e. in the rubbery state where segment mobility is important. Considering that isomerization occurs in the rubbery state at a rate similar to that in solution, it should be expected that the use of isomerization reactions for photocontractile behavior should be optimal with easily deformable networks, i.e. swollen gels and rubbery materials.

It was first stated by Lovrien[16] that, if a macromolecule is involved in equilibrium interactions with some photoisomerizable species present in solution, it may be forced to assume a different conformation under the influence of light and conversion of light energy into chemical energy may result. Likewise, if a macromolecule containing photo-isomerizable cross-links changes dimensions on irradiation, conversion of light energy to mechanical energy may occur. Thus, Lovrien reported the dependence of the conformational behavior of synthetic polyelectrolytes on the cis-trans ratio of an absorbed azo dye with the resulting change of the solution viscosity on irradiation.

It is on the basis of Lovrien's ideas concerning the photoregulation of polymer conformation in solution by means of photochromic species that Van der Veen & Prins[54] reported a first photomechanochemical model transducer, consisting of water-swollen membranes of poly(2-hydroxyethyl methacrylate) cross-linked with ethylene glycol dimethacrylate (1.1 wt-%) in the presence of a sulfonated bis-azostilbene dye (chrysophenine G), the ratio chrysophenine/2-hydroxyethyl methacrylate being 1/400.

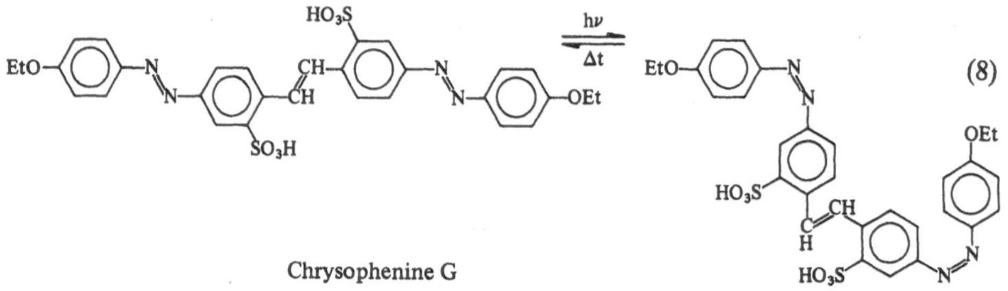

Chrysophenine G

(8)

On irradiation the t-t-t-dye is isomerized to its c-t-c-isomer: this conformational change causes a decrease of dye-polymer interactions resulting in a gel contraction of 1.2%, i.e. a change in volume of about 3.6%. In the dark the gel recovers its original dimensions; the rate of recovery is a function of temperature and parallels the increase of optical density of chrysophenine at 400 nm.

Later, they reported[55, 56] the photoregulation of the degree of ionization and swelling of poly(methacrylic acid) membrane cross-linked with 1 mol-% ethylene glycol dimethacrylate onto which positively charged p-phenylazophenyltrimethylammonium ions (PTA) were absorbed (Eq. (9)).

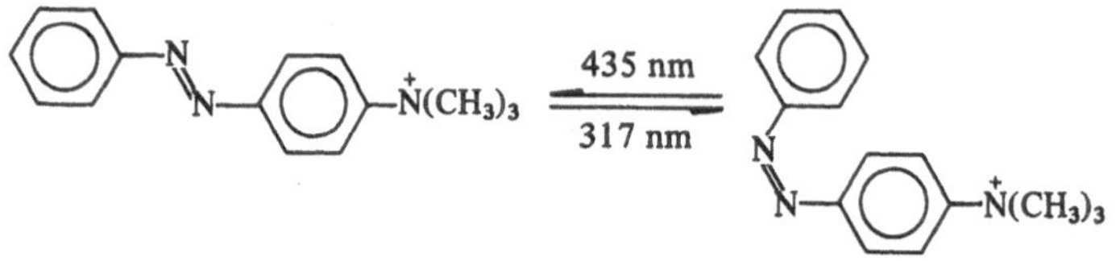

trans–PTA   cis–PTA

(9)

Photoisomerization of trans PTA to cis PTA causes a change of the degree of ionization (by half a pK unit) and thus of the degree of swelling.

Changes in the dimensions of a polyelectrolyte gel may also result from a chemically induced ionization irrespective of isomerization. If this ionization is caused by irradiation with light and if the lifetime of the charged species is sufficiently long to permit the polymer to deform, a mechanophotochemical effect may result[57]. This effect was described by Aviram in the case of poly[p-(N,N-dimethylamino)-N-γ-D-glutamanilide] cross-linked with 1.5% 2,6-bis-(bromomethyl)naphthalene; on irradiation in the presence of carbon tetrabromide as an acceptor photoionization occurs (Eq. (10)).

In solution the conductivity is strongly increased by the photochemical generation of the ions; films display up to 35% dilation in each dimension when exposed to light. It should however be noted that no attempts were made to reverse the dilated gels back to their original size.

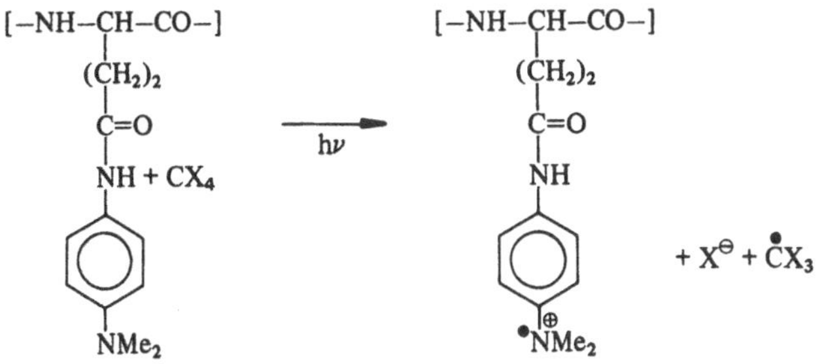

[−NH−CH−CO−]   [−NH−CH−CO−]
    | |
  (CH₂)₂   (CH₂)₂
    | |
  C=O $\xrightarrow{h\nu}$ C=O
    | |
 NH + CX₄   NH   (10)

    ⬡   ⬡ $+ X^{\ominus} + \overset{\bullet}{C}X_3$

 NMe₂   •ⁿNMe₂⁺

Smets and coworkers[44, 46, 58, 59] worked with stretched spirobenzopyran rubber networks obtained by copolymerization of ethyl acrylate with variable amounts of a bis-photochrome dimethacrylate as cross-linking agent, namely 1,1'-(α,α'-p-xylyl)-bis-[3',3'-dimethyl-8-methacryloyloxymethyl-6-nitro-spiro(2 H-1-benzopyran-2,2'-indoline)]. The chemical structure of these DIPS-rubbers is given in Fig. 11.

The photomechanical behavior of the networks was studied under constant stress (load) by following at constant temperature the length contraction as a function of the period of irradiation; in the dark, length recovery takes place although at a lower rate

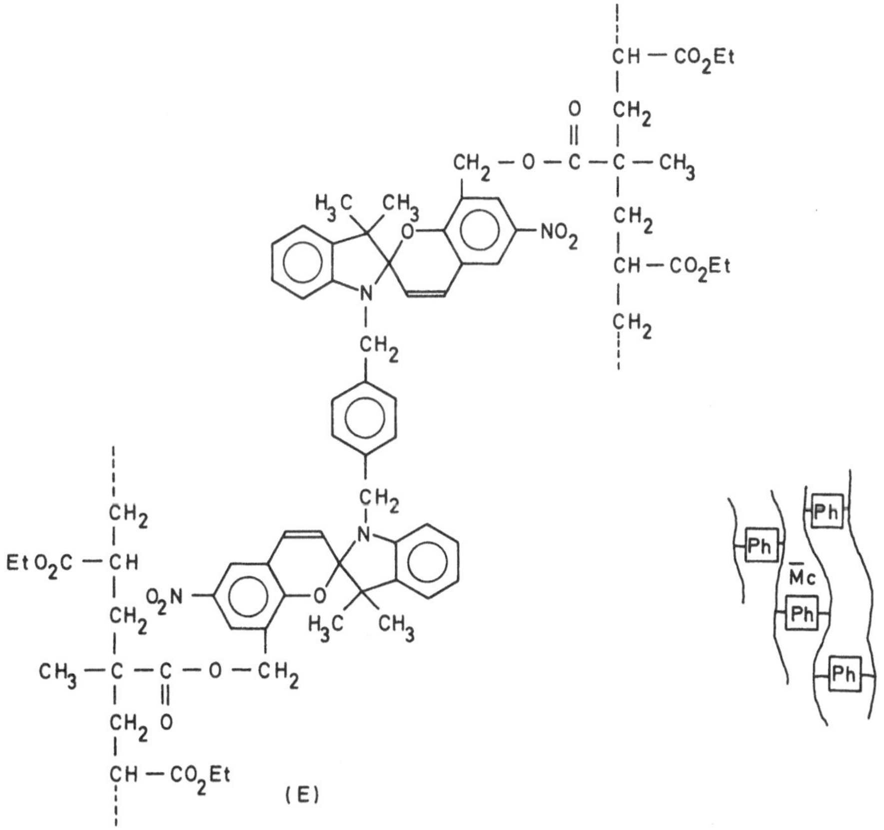

Fig. 11. Photochromic poly(ethyl acrylate) rubber cross-linked with bis-photochrome dimethacrylate

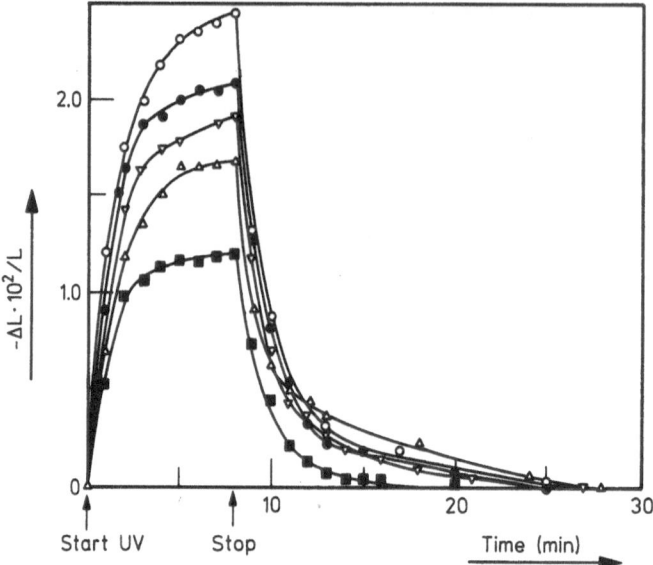

Fig. 12. Photomechanical behavior of ethyl acrylate-bis-(DIPS-methacrylate) copolymers ($T_g = -15\ °C$). Influence of temperature and stress time in minutes

	Temperature	mm	load
○	15	44.5	21.9
△	29.6	44	21.9
■	45	42	21.9
●	15	22	35.7
▽	15	32.5	59.4

than shrinkage. This photomechanical effect increases with decreasing temperature and decreasing stress; the light/dark cycles can be repeated several times and are nicely reversible. Figure 12 illustrates the typical photocontractile behavior of a copolymer containing 0.5 mol-% bis (DIPS methacrylate).

An optimum stress at which a maximum photocontraction takes place (2-3% relative contraction) exists; it depends on the degree of cross-linking of the network (Fig. 13).

Initially[59], it was assumed that the contraction corresponds to an entropy increase of the polymer chain due to the higher flexibility of the open-ring merocyanine compared to the stiffness of the parent ring-closed spiropyran.

Following, however, simultaneously color fading and length recovery in the dark[44], only a minor decrease (<3%) of the optical density for complete length recovery was observed, and the rate of contraction is higher when the film has previously been coloured than on the first irradiation cycle. Moreover, dark decoloration kinetics indicates an activation energy of 26.8 kcal/mol, while length recovery rate measurements give only $E_a = 6–7$ kcal/mol.

It is obvious that such a difference in the activation energy of both phenomena indicates that fading of merocyanine to spirobenzopyran cannot be the determining

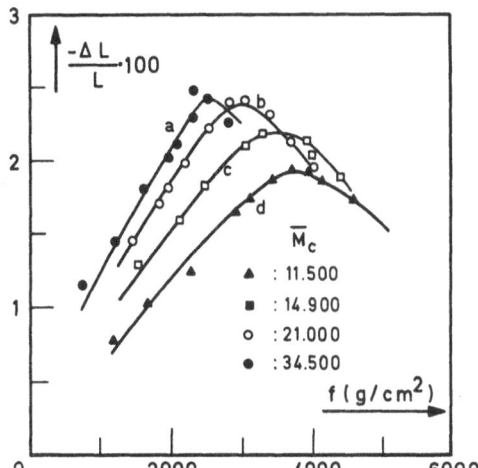

Fig. 13. Influence of tensile force f and M_c on the photocontraction of photochromic poly(ethyl acrylate) rubbers M_c = mean molecular weight between two next cross-links

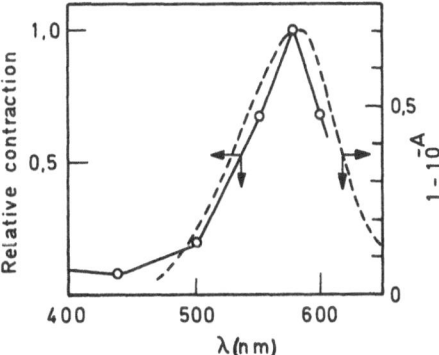

Fig. 14. Contraction (——) and absorption (– – –) spectra of photochromic poly(ethyl acrylate) rubbers

factor for length recovery except for the first formation of merocyanine. By using several interference filters the dependence of film shrinkage on the wavelength of irradiation was determined, taking into account the differences of light intensity at the different wavelengths. The contraction spectrum fits very well the visible light absorption spectrum of merocyanine, as can be seen in Fig. 14.

On the other hand, the contraction efficiency of ultraviolet light is poor, notwithstanding the strong absorption of the spirobenzopyrans at 367 nm. The interpretation of these results can be based on:

i) interconversion of the different merocyanine isomers[30, 31] in the solid polymer. During light absorption the most stable isomer is converted into another one. This may result in a change of the conformation of the neighboring chain segments which may be responsible for the observed contraction. On cutting off light, the stereoisomer equilibrium is restored progressively and the most stable isomer reformed with concomitant length recovery. Such interpretation is compatible with the low energy of activation; it is also supported by a change in the shape of the absorption band of decoloration with time (λ_{max} progressively from 586 to 562 nm).

ii) · Local temperature increase resulting from light absorption which would also cause shrinkage of the stretched rubber. As pointed out previously, the absorbed light energy exceeds considerably the activation energy of ring opening, i.e. the excess energy is transformed into heat, transmitted to the surroundings, and may cause conformational changes. Such thermal effects are presumed to occur notwithstanding the temperature-controlled set-up; their contribution to the contraction should increase with the duration and intensity of irradiation and with the content of chromophore.

It should be noted, however, that in the above experiments, the samples contain only 0.44 to 0.76 mol-% chromophore and that the rapid contraction phase, which amounts to 80% of the final contraction, lasts only for 20–30 s irradiation using interference filters.

iii) Isomerization of a small fraction of the spirobenzopyran rings which are present in locally strained sites in the polymeric matrix. This interpretation is similar to that given above for the photochromic behavior of spiropyran in bulk polymers and may very well explain the photomechanical effects in polyamides and polyimides described in the next paragraphs. For soft poly(ethyl acrylate)rubbers ($T_g \sim -15\ °C$) it seems however less acceptable.

It is therefore admitted that the photochemically induced contraction/length recovery phenomena are related mainly to the photochemical interconversion of the merocyanine isomers, and at a smaller extent to local thermal effects (20–30%).

Noteworthy is the efficiency of energy conversion of these spirobenzopyran rubbers ϕ_E, i.e. the ratio of the contraction work to the energy of absorbed light. For irradiation periods of 60 s the mean value of ϕ_E is 0.86% independent of the irradiation wavelength between 500 and 600 nm. Calculated on the initial rapid contraction phase it is equal to 2%.

Similar reversible contraction/dilation experiments under constant load were recently performed by Eisenbach[60] on stretched poly(ethyl acrylate) networks, cross-linked however with 4,4′-dimethacryloylaminoazobenzene (0.02 mol-%).

$$CH_2{=}C(CH_3){-}CONH{-}\bigcirc{-}N{=}N{-}\bigcirc{-}NHCO{-}C(CH_3){=}CH_2$$

XI

Upon irradiation trans → cis isomerization causes conformational changes of adjacent network segments which are considered to be responsible for the photomechanical effect. The observed contraction is however small and amounts only to about 0.15–0.25%.

In the poly(ethyl acrylate)rubbers described above spirobenzopyran and aromatic azo chromophores were incorporated in the cross-links between the polyacrylic chains. In contrast, Matejika and al.[61, 62] studied cross-linked systems with azo side groups; they considered especially the relative importance of the thermal effects occurring during photomechanical conversion. Most interesting are their results on copolymers containing low concentrations of chromophore (for which heat effects are less important), namely copolymers of β-hydroxyethyl methacrylate (0.99) and azonaphthol methacrylate (XII) (0.01) swollen in water

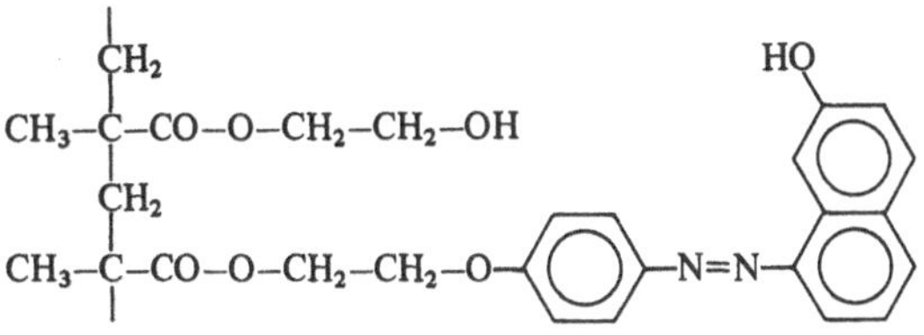

CH$_2$
|
CH$_3$–C–CO–O–CH$_2$–CH$_2$–OH
|
CH$_2$
|
CH$_3$–C–CO–O–CH$_2$–CH$_2$–O⟨benzene ring⟩–N=N–⟨naphthalene ring with HO⟩
|

XII

and copolymers of butyl acrylate (0.946) and methacrylamido-azobenzene (XIII) (0.054) examined in dry state.

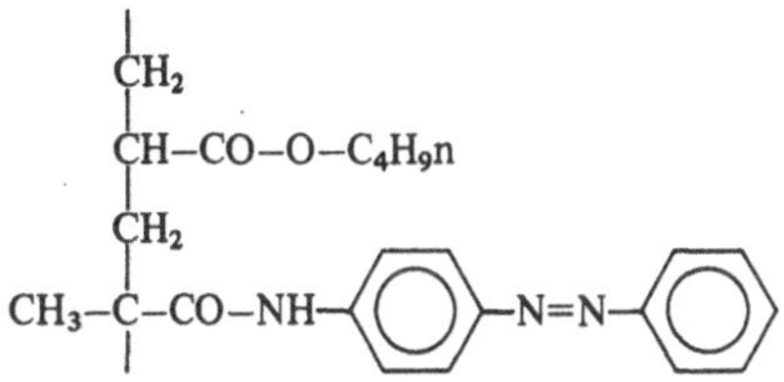

CH$_2$
|
CH–CO–O–C$_4$H$_9$n
|
CH$_2$
|
CH$_3$–C–CO–NH–⟨benzene ring⟩–N=N–⟨benzene ring⟩
|

XIII

Both copolymers were cross-linked by copolymerization with (1–2%) ethylene dimethacrylate; the effect of irradiation was followed by measuring the change of elastic retractive force at constant elongation. In the first case the contraction upon irradiation of the gel is mainly due to changes of chain conformation and swelling equilibrium induced by trans-cis photoisomerization. In the second case by comparing the retractive force for irradiated and unirradiated samples the heat effect was evaluated, and the photoinduced contraction estimated to 1%.

As could be expected from molecular model considerations, a comparison of the available data shows that the spirobenzopyran chromophores cause more important conformational changes of neighboring chain segments and consequently stronger photo-contractile effects than the aromatic azo chromophores.

3.2 Uncross-Linked Bulk Polymers

Besides the photomechanical effects observed on rubbery networks and swollen gels, photocontractility of photochromic systems in the solid state without chemical cross-linking agent has been described. In some cases, possibly partial crystallinity and eventually hydrogen bonding ensure physical cross-linking and consequently a network structure. In most simple cases, the photosensitive compounds were dissolved in a polymeric matrix, and the effects are based on interactions between the dye and polymer substrate molecule in the same sense as suggested by Lovrien[16] and Prins[54–56]. The first observation of photocontractile effects was that of Merian and al.[63] on cellulose acetate ribbon dyed with the isomerizable azo dye (XIV)

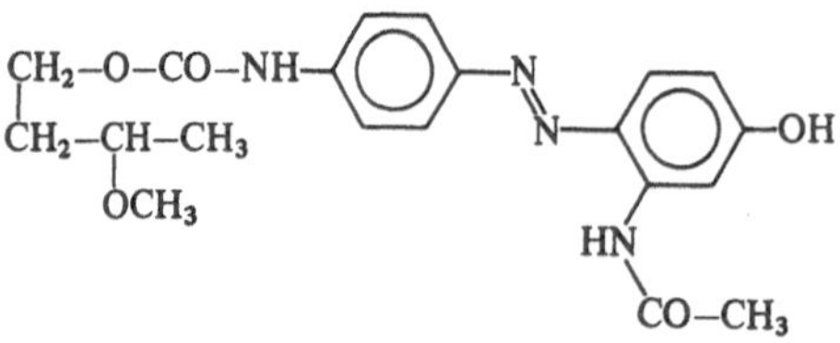

XIV

On exposure to sunlight the sample contracted and showed a pronounced change of shade to orange; on storage in the dark, it recovers its original length and shade. The effect was reproducible several times, but only in the presence of an isomerizable chromophore.

Blair and Law[64] described photoresponsive effects under constant load of nylon 6,6 films dyed with trans β-carotene and with trans 4-nitro-4'-hydroxy-α-cyanostilbene the contractions of which amount to 0.6 and 0.8% respectively. In the absence of a dye no photomechanical effect occurs. For the polyamide/trans-β-carotene system length recovery is very slow and reversible incompletely; it parallels, however, the change in absorption of trans-β-carotene. Very recently, Blair and Pogue[65] described photomechanical effects occurring in polystyrene and poly(methyl methacrylate)films containing 6'-nitro-1,3,3-trimethyl-spiro-(2'H-1'-benzopyran-2,2'-indoline), i.e. 1-methyl-6'-nitro DIPS. In both systems, the effects in terms of change in length were small, but easily followed by measuring stress changes with time at constant length. The photomechanical effect depends on the concentration of spirobenzopyran; in polystyrene films it increases rapidly with increasing concentration, passes through a maximum at 5 wt-% and then decreases slowly. The authors suggest that the photomechanical effect is due to the isomerization of only those molecules which take part in the initial fast decoloration reaction (see Chap. 2.1.2); these molecules represent only a small fraction of the photochrome molecules and occur in a strained state within a restricted location. Noteworthy is that the effect obtained with these last systems is opposite to that described above concerning photochromic networks where the chromophores were incorporated into the polymer.

When studying systems in which the photochromes are directly bound to the polymers, first mention should be made of the report of Agolini and Gay on the photo- and thermocontractile behavior of azoaromatic polyimides (XV) obtained by condensation of 4,4'-diaminoazobenzene with pyromellitic anhydride and dehydration of the resulting polyamine acid[66].

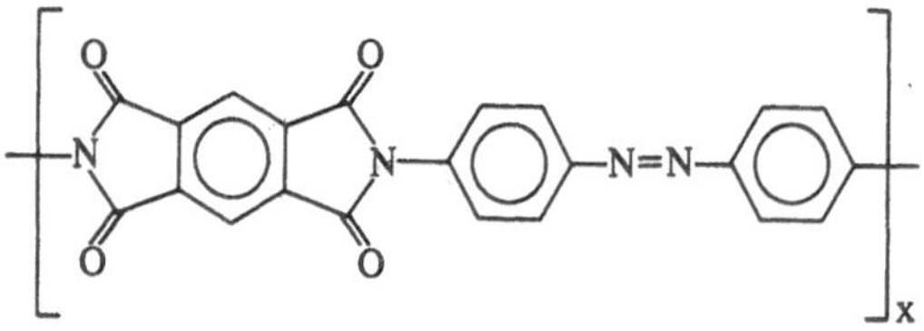

XV

Since the polymer is semicrystalline, trans-cis azo isomerization must be restricted to the amorphous regions. Measurements at constant stress carried out at 200 °C indicate a deformation of about 0.6%. On the other hand, in experiments at constant length the stress imposed on a film during irradiation increases with time, the effect being small but real. Though the contraction is associated with trans → cis conversion it shows only little or no activation energy, in contrast to the activation energy of 22 kcal of the isomerization of azo compounds. It seems therefore that the rate of dilation/contraction is controlled by some viscoelastic properties of the polymer (with low E_a) rather than by the rate of isomerization of the azo group.

Blair, Pogue and Riordan[17] described photoresponsive effects of photochromic polyamides in which every monomer unit contains an azo group, namely the 3,3'-azodibenzoyl-trans-3,5-dimethylpiperazene (XVI) and its 4,4'-isomer (XVII):

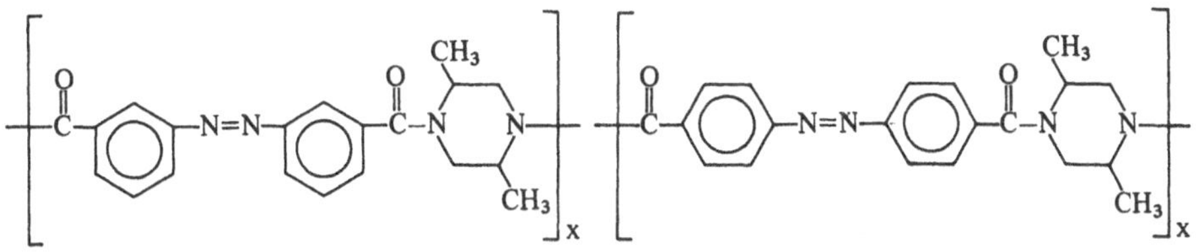

XVI   XVII

On irradiation the measured stress increases indicating a contraction of the sample up to a photostationary state; in the dark the stress decreases again, and the cycle can be repeated many times. Here again the relaxation rate in the dark is much more rapid than the normal cis-trans return in solution, and should be due to the isomerization of a small fraction of azo residues in local strained non-equilibrium sites in the polymer matrix.

From a similar point of view Osada and Katsumura[21] examined the photomechanical energy conversion of a polyamide containing a stilbene chromophore in the backbone (XVIII):

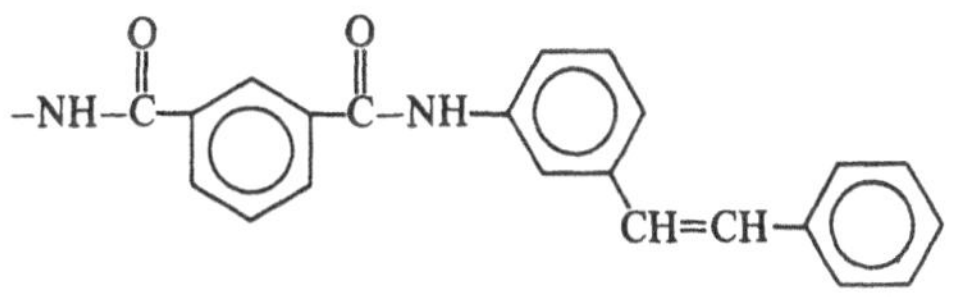

XVIII

The polyamide was obtained by condensation of 3,4'-diaminostilbene with isophthaloyl chloride. It contains, as in the preceding case, an isomerizable stilbene group in every unit. Measuring the stress of the polymer film during irradiation with a strain gauge they found unexpectedly that the film, relaxed on irradiation, assumes the initial values of stress in the dark. In contrast, in solution, the viscosity decreases gradually on irradiation. Such apparent experimental contradiction means that, besides conformational changes occurring in solution, neighboring segment mobility has to be considered in the solid state and may become the predominant factor for bulk polymers.

3.3 Photoresponsive Polymer Monolayers

The ring opening isomerization of twisted spirobenzopyran to planar merocyanine on irradiation with UV light may cause a reduction of apparent area of the spiran molecule in polymer monolayers. Thus Blair and Pogue[67] examined mixed monolayers containing various amounts (10 to 40 p/w photochrome/polymer) of 6'-nitro-DIPS in poly(methyl methacrylate) using a Langmuir film balance. Indeed, they measured a reduction in the apparent area of the spiran molecule exposed to UV irradiation. In this case where the chromophores are dissolved in the macromolecules, other factors also intervene, e.g. miscibility of the components, spreading nature of the mixed monolayer, zwitterionic structure of the merocyanine. These difficulties were avoided by Rondelez and al.[68] who studied the photoinduced mechanical response of a monolayer consisting of a copolymer of methyl methacrylate and 1-(β-methacryloyloxyethyl)-6'-nitro-DIPS (M = 215000) at an air-water interface. Irradiation with ultraviolet light pulses leads, at the saturation state, to an increase of the surface pressure of about 10%. The change in pressure decreases progressively in the dark and vanishes after about 20 min: this decrease can be considerably accelerated by the use of visible light pulses. The process can be repeated several times and shows that the photoinduced mechanical response is fully reversible (Fig. 15).

Very recently, Rondelez[69] investigated the influence of the pH of the water substrate on the photoinduced surface pressure changes. These changes are larger at low pH at which complete reversibility of the photochemical strain can be observed on excitation with visible light. At high pH the merocyanine should be momentarily photoconverted to a different isomer but no return to the initial value has been detected.

Blair et al.[17] have also carried out Langmuir film balance measurements on monolayers of the photochromic polyamides mentioned above, namely the 3,3'- and 4,4'-poly-(azodibenzoyl-trans-2,5-dimethylpiperazene) (XVI and XVII). Changing from dark to light a reduction in the area per monomer unit was observed and interpreted as resulting from azo trans to cis isomerization. The differences between the pressure-area curves of the meta(3,3'-) and para (4,4'-)polyamides were interpreted on the assumption that the meta compound has a helical structure and that the para compound is linear.

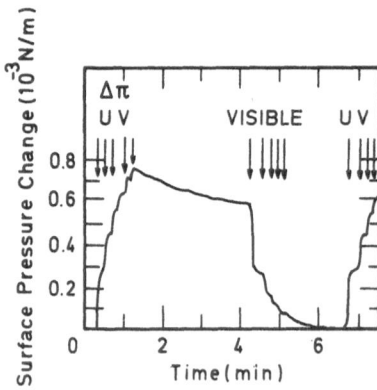

Fig. 15. Surface pressure of photochromic copolymer monolayer vs. time upon irradiation with UV and subsequently with visible light. Physic. Rev. Letter. 44, 591 (1980)

It is obvious, as stated by Rondelez[68], that these photochromic phenomena occurring in two dimensions result in a dynamic photoresponse of polymer films. These films can be considered as model photoreceptors for photoregulated biological processes; they could play an important role in many fields as photoprinting and photocontrolled drug release from vesicles.

Acknowledgement. The author acknowledges gratefully that much of his research and of his coworkers on photochromic materials has been supported by the Ministry of Scientific Programmation and by the Institute for Research in Industry and Agriculture, Belgium.

4 References

1. Irie, M., Hayashi, K.: Photochromic phenomena in liquid phase. Adv. Polym. Sci. Berlin, Heidelberg, New York: Springer Verlag, in press
2. Kuhn W., Künzle, O., Katchalsky, A.: Bull. Soc. Chim. Belge *57*, 421 (1948)
3. Katchalsky, A., Künzle, O., Kuhn, W.: J. Polym. Sci. *5*, 283 (1950)
4. Künzle, O.: Rec. Trav. Chim. Pays-Bas *67*, 699 (1949)
5. Kuhn, W., Katchalsky, A., Eisenberg, H.: Nature *165*, 514 (1950)
6. Katchalsky, A., Eisenberg, H.: Nature *166*, 267 (1-50
7. Osada, Y., Y. Saito: Makromol. Chem. *176*, 2761 (1975)
8. Kun, W.: Angew. Chem. *70*, 58 (1958)
9. Kuhn, W., Ramel, A., Walters, D. H.: Angew. Chem. *70*, 314 (1958)
10. Kuhn, W., Ramel, A., Walters, D. H.: Size and shape changes of contractile polymers. Conversion of chemical into mechanical energy. Wasserman, A. (ed.), p. 41. New York: Pergamon Press 1960
11. Katchalsky, A., Zwick, M.: J. Polym. Sci. *16*, 221 (1955)
12. Kuhn, W., Toth, I., Kuhn, H. J.: Makromol. Chem. *60*, 77 (1963)
13. Osada, Y., Sato, M.: Polymer *21*, 1061 (1980)
14. Osada, Y.: J. Polym. Sci. Polym. Chem. Ed. *15*, 255 (1977)
15. Osada, Y.: J. Polym. Sci. Polym. Lett. Ed. *18*, 281 (1980)
16. Lovrien, R.: Proc. Nat. Acad. Sci. U.S. *57*, 236 (1967)
17. Blair, H. S., Pogue, H. I., Riordan, E.: Polymer *21*, 1195 (1980)
18. Irie, M. et al.: Macromolecules *14*, 262 (1981)
19. Irie, M., Schnabel, W.: Macromolecules *14*, 1246 (1981)
20. Matějka, L., Dušek, K.: Makromol. Chem. *182*, 3223 (1981)
21. Osada, Y., Katsumura, E.: Makromol. Chem. Rapid Commun. *2*, 241 (1981); *2*, 411 (1981)
22. Irie, M. et al.: J. Polym. Sci. Polym. Lett. *17*, 29 (1979)
23. Irie, M. Menju, A., Hayashi, K.: Macromolecules *12*, 1176 (1979)
24. Menju, A., Hayashi, K., Irie, M.: Macromolecules *14*, 755 (1981)
25. Kamogawa, H., Kato, M., Sugiyama, H.: J. Polym. Sci. A-1, *6*, 2967 (1968)
26. Gardlund, Z. G.: J. Polym. Sci., *6*, 57 (1968)
27. Gardlund, Z. G., Laverty, J. J.: J. Polym. Sci., Polym. Lett. *7*, 719 (1969)
28. Smets, G.: Pure Appl. Chem. *30*, 1 (1972)
29. Verborgt, J., Smets, G.: J. Polym. Sci., Polym. Chem. Ed. *12*, 2511 (1974)
30. Chaudé, O.: Cah. Phys. *52*, 39 (1954)
31. Arnaud, J., Wippler, C., Beau d'Angères, F.: J. Chim. Phys. *64*, 1165 (1967)
32. Eisenbach, C.: Makromol. Chem. *179*, 2489 (1978)
33. Eisenbach, C.: Polymer Bulletin *1*, 517 (1979)
34. Paik, C. S., Morawetz, H.: Macromolecules *5*, 171 (1972)
35. Vandewyer, P., Smets, G.: J. Polym. Sci. *C 22*, 231 (1968)
36. Smets, G.: IUPAC Internat. Symp. Macromolecules, Budapest 1969, pp. 65–87
37. Miura, M. et al.: Polymer *19*, 348 (1978)

38. Heiligman-Rim, R., Hirshberg, Y., Fischer, E.: J. Phys. Chem. *66*, 2465 (1962)
39. Kryszewski, M., Nadolski, B.: J. Polym. Sci., Polym. Chem. Ed. *13*, 345 (1975)
40. Kryszewski, M., Nadolski, B.: Pure Appl. Chem. *49*, 511 (1977)
41. Lawrie, N. C., North, A. M.: European Polym. J. *9*, 348 (1973)
42. Kryszewski, M., Grachowska-Kapienis, D., Nadolski, B.: J. Polym. Sci., Polymer Chem. Ed. *11*, 2423 (1973)
43. Smets, G., Thoen, J., Aerts, A.: J. Polym. Sci., Polym. Symp. *51*, 119 (1975)
44. Smets, G., Braeken, J., Irie, M.: Pure Appl. Chem. *50*, 845 (1978)
45. Flannery, J. B. Jr.: J. Amer. Chem. Soc. *90*, 5660 (1968)
46. Smets, G., Evens, G.: Pure Appl. Chem. Suppl. Macromol. Chem. *8*, 357 (1973)
47. Eirich, F. R.: Rheology, theory and applications, Vol. I, p. 441–52. New York: Academic Press 1956
48. Williams, M. L., Landel, R. L., Ferry, J. D.: J. Amer. Chem. Soc. *37*, 3701 (1955)
49. Sevens, G.: Ph. D. thesis K. Universiteit Leuven, pp. 52–60, 1972
50. Eisenbach, C.: Ber. Bunsenges. Phys. Chem. *84*, 680–90 (1980)
51. Eisenbach, C.: Photogr. Sci. Eng. *23*, 183 (1979)
52. Eisenbach, C.: Polym. Bull. *2*, 169 (1980)
53. Sung, C. S. P., Lamarre, L., Chung, K. H.: Macromolecules *14*, 1839 (1981)
54. Van der Veen, G., Prins, W.: Nat. Phys. Sci. *230*, 70 (1971)
55. Chuang, J. C., De Sorgo, M., Prins, W.: J. Mechanochem. Cell Mobility *2*, 105 (1973)
56. Van der Veen, G., Hoguet, R., Prins, W.: Photochem. Photobiol. *19*, 197 (1974)
57. Aviram, A.: Macromolecules *11*, 1275 (1978)
58. Smets, G.: Pure Appl. Chem. *30*, 1 (1972)
59. Smets, G., De Blauwe, F.: Pure Appl. Chem. *39*, 225 (1974)
60. Eisenbach, C. D.: Polymer *21*, 1175 (1980)
61. Matějika, L., Dušek, K., Ilavský, M.: Polym. Bulletin *1*, 659 (1979)
62. Matějika, L. et al.: Polymer *22*, 1911 (1981)
63. Husy, H., Merian, E., Schetty, G.: Textile Res. J. *39*, 94 (1969)
64. Blair, H. S., Kau Law, T.: Polymer *21*, 1475 (1980)
65. Blair, H. S., Pogue, H. I.: Polymer *23*, 779 (1982)
66. Agolini, F., Gay, F. P.: Macromolecules *3*, 349 (1970)
67. Blair, H. S., Pogue, H. I.: Polymer *20*, 99 (1979)
68. Gruler, H., Vilanove, R., Rondelez, F.: Phys. Rev. Lett. *44*, 590 (1980)
69. Vilanove, R. et al.: Macromolecules, in press (1982)

Received August 4, 1982
S. Okamura (Editor)

Polymer Square Planar Metal Chelates for Science and Industry.

Synthesis, Properties and Applications

Dieter Wöhrle

Organic and Macromolecular Chemistry, Bremen University, Bibliotheksstr. NW 2, D-2800 Bremen, Federal Republic of Germany

The topic of this review concentrates on those metal chelates where the metal atom is surrounded by a planar ligand: porphyrins, phthalocyanines, hemiporphyrazines (hexaazocyclanes), tetraaza(14)-annulenes, bis(1,2-dioximes), and Schiff base chelates (salens). The coordinative (through the metal atom) and covalent (through the ligand and metal atom) binding of the chelate in a polymer chain is described in detail. Binding of small molecules as oxygen and catalytic activity are the main promising properties. Then the covalent and coordinative incorporation of the chelate through the metal atom in the polymer chain is discussed. Electrical conductivity and thermal properties are further main interesting aspects of such polymers. Finally, this review deals with polymeric chelates with covalent introduction of the chelate through the ligand in the main chain. These polymeric chelates and their ligands show other interesting features, such as electrical conductivity, small molecule binding, catalytic and electrocatalytic properties.

Advances in Polymer Science 50
© Springer-Verlag Berlin Heidelberg 1983

1 Introduction. Selection of Metal Chelates

The combination of metal ion, ligand and chemical environment (such as solvent or polymer) determines the chemical and physical properties of the metal chelates. Biological metal porphyrins occuring in hemoglobin, chlorophyll, vitamin B_{12} and some metalloenzymes show this extremly well. Model systems seems to be useful in order to elucidate these factors and to construct artificial systems for practical use.

Knowledge about natural metal chelates induced in the first stage research work for artificial low molecular metal chelates. Secondly experience about the important environmental effect of natural polymers like apoprotein lead to research on the combination of synthetic polymers with a metal chelate. By intensive investigation in these fields new combinations of metal chelate/polymer may be found, even with unknown properties.

This review deals with synthetic polymer bond and polymer metal chelates like porphyrins and others containing a square planar ligand. Generally, four N-(N_4-chelates) or two N-, two O- (N_2O_2-chelates) ligand atoms are surrounding the metal atom in square planar arrangement symbolized through structure (A) and (B).

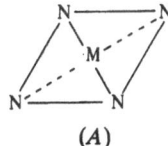

(A)

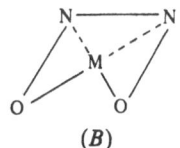

(B)

Examples for low molecular metal chelates included in this report are:
- N_4-chelates
  - porphyrins, P (1)[2-5]
  - phthalocyanines, PC (2)[1, 5-8]
  - hemiporphyrazines, HP (3)[1, 9, 10]
  - tetraaza(14)annulenes, TAA (4)[11]
  - bis(1,2-dioximes) (5)[12]
- N_2O_2-chelates
  - Schiffbase chelates (6)[13-15]

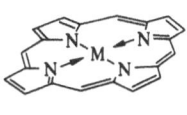

(1), P

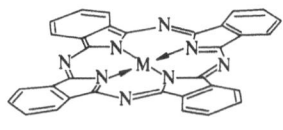

(2), Pc

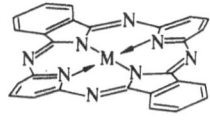

(3), Hp

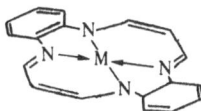

(4), Taa

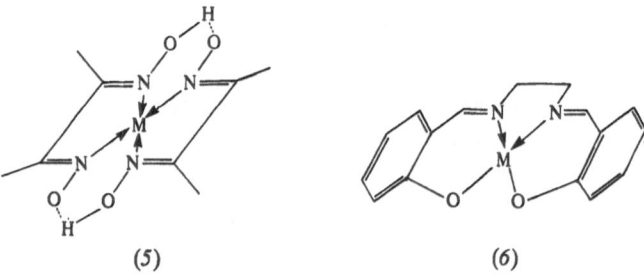

(5)   (6)

In every case the ligand is dianionic. With a bivalent metal ion a neutral metal chelate is obtained. The metal free ligands contain two hydrogens instead of the metal. The structure of the chelates are well investigated showing the positions of the 4 ligand atoms at the edges of a square. The synthesis of the low molecular chelates is described elsewhere[1-15].

The following possibilities have the metal chelates as part of a polymer.

Chapter 2: Coordinative polymer bond in metal chelates

A polymer with ligand groups is coordinated in axial position to the metal chelate with coordinative bond. When starting from a square planar chelate five or six coordination places at axial positions lead to pyramidal or octahedral surrounding of the metal (structures C, D).

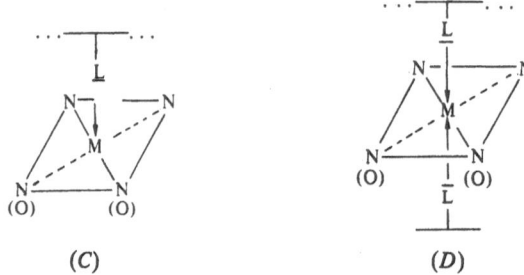

(C)   (D)

Such a bond is realized with polymer ligands containing basic groups with σ-donor and π-acceptor properties like pyridine, imidazole. Coordination chemistry is nearly equal to low molecular axial ligands.

Chapter 3: Covalent polymer bound metal chelates

As shown in structure (E) a covalent bond is existing between the polymer backbone and the ligand of the metal chelate. If the polymer contains basic donor groups additional coordinative bonds may exist axial as mentioned in Chap. 2. Also a covalent bond from the polymer to metal atom of the chelate may occur (structure F).

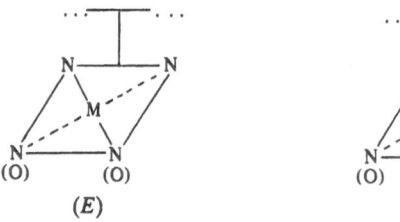

(E)   (F)

Chapter 4: *Chelates with the central metal atom in the polymeric chain*

In this case the metal atom is a part of the polymer chain (structure *G*). To realize such an arrangement with the mentioned ligands four-or higher valent metal atoms like Si, Se, Sn and other may be taken.

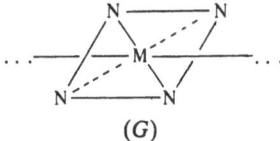

(*G*)

Chapter 5: *Polymeric metal chelates through the ligand*

The ligand as shown in structure (*H, I*) is now part of the polymer chain or network, e.g. polymer phthalocyanines. The best way for their synthesis is starting from a tetracyano compound as mono-mer. So the reactivity of the nitrile-group forms the polymer with (C=N)-bonds. After a review on "Polymerisation of some nitriles"[1] newer literature is only taken into account.

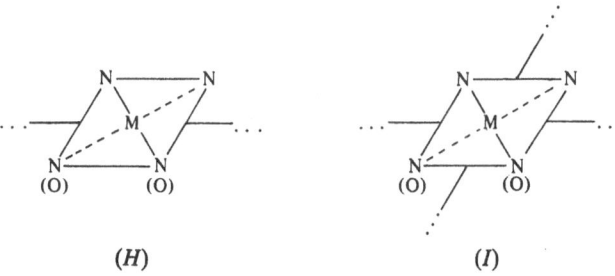

(*H*)   (*I*)

The following properties of the polymer chelates in each specific chapter are described:
- binding of small molecules like oxygen
- catalytic properties
- electrocatalysis like fuel cell
- light energy conversion like photoredoxreaction
- electrical conductivity
- thermal stability

2 Coordinative Polymer Bond in Metal Chelates

Natural proteides like hemoglobin, myoglobin, peroxidase, catalase, cytochrom a, b, P 450, desoxigenase and chlorophyll contain the porphyrin system as a prostetic group. The porphyrin is bound coordinatively through the metal atom to the natural polymer or inserted in lipid-protein layers. Important properties are binding of small molecules, catalysis and electron transfer reactions.

For *constructing an artificial system* to conserve chelate properties we must under-stand the natural system. The main functions of hemoglobin (with heme (*7 a*) as prostetic group, which is most examined) are (Scheme 1):

An active metal ion like Fe(II) is obtained by surrounding it with a square planar porphyrin system
$\quad$ (a → b)

- An organic donor ligand (b → c → d) like imidazole working as σ-donor and π-acceptor is the thermodynamic condition for binding and activating small molecules like O_2 (c → e)
- Free Fe(II)-porphyrins are easily oxidized at room temperature through peroxobridged (e → f) to μ-oxo-bridged *dimers* (f → g) which are generally not working reversibly. So a polymer like natural globin inhibits dimerisation
- The globin pocket pressents *dissoziation* of the porpyrin from the ligand
- Intercalation of Fe(II)-porphyrin in the *hydrophobic pocket* of globin inhibits oxidation of mononuclear oxygen adduct to Fe(III)–OH (e → h) and hydroperoxide ion formation
- *Hydrophylic periphery* of the globin leads to water solubility
- Allosteric effect of the hemoglobin during oxygen uptake

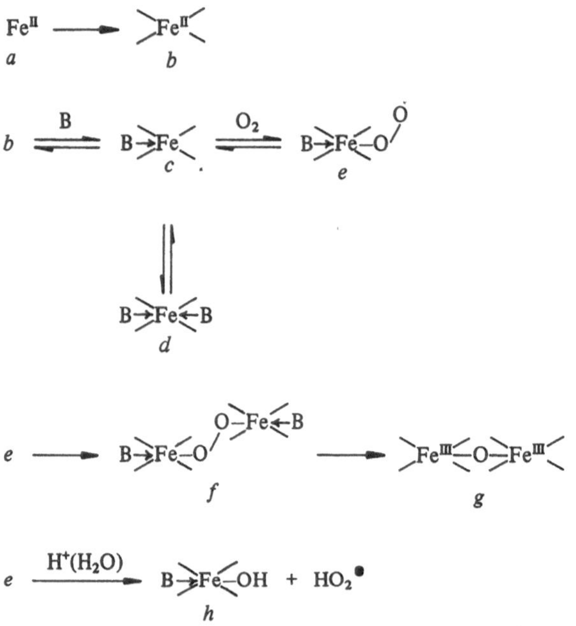

Scheme 1

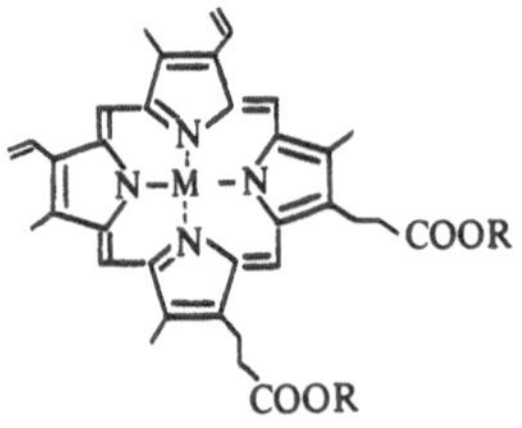

(7)	M	R
a	Fe^{2+}	-H
b	$Fe^{3+}Cl^-$	-H
c	Fe^{2+}	-alkyl
d	$Fe^{3+}Cl^-$	-alkyl
e	Co^{2+}	-alkyl
f	$2H^+$	-alkyl
g	Co^{2+}	-H
h	$2H^+$	-H

Considering the synthesis of an artificial system with coordinative or covalent bonds in square planar metal chelates for binding small molecules or catalytic use, main points are donor ligand activation, reduced aggregation, and irreversible oxidation of the chelate.

The advantage of the coordinative polymer binding (structures C, D) is the easy way of preparation. Normally, only solutions of polymer ligand and low molecular metal chelate are mixed (Scheme 2).

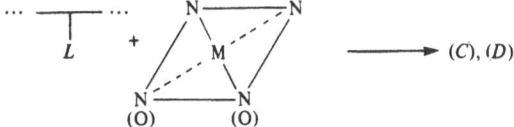

Scheme 2

2.1 Porphyrins (*1*)

2.1.1 Synthesis

In 1977 Tsuchida and Nishide presented a review on polymer metal complexes also containing porphyrins[16]. Therefore only newer results are discussed and compared with former research work.

Low molecular porphyrins from natural sources and synthetic procedures are used for the combination with polymer ligands:

- Fe(II, III)-protoporphyrin-IX (heme, hemin) (*7a, b*) with poly(L-lysine)[17–22], poly(L-histidine)[23], poly(γ-benzyl-L-glutamate (with pendant imidazole)[19, 24], polyethylenimine[19, 22], poly(4-vinyl-pyridine) (also partly quarternized)[19, 22, 25–31, 32], poly(N- or 4(5)-vinylimidazoles) (partly substituted at position 2)[19, 25, 33–37], water soluble imidazole modified polyphophazenes[38], macroporous methyl methacrylate, with covalent bound imidazole[39, 40].
- Fe(II, III)-protoporphyrin-IX-diester (*7c, d*) with poly(N-vinyl)-imidazole)[41, 42], polystyrene (in mixture with 1-(2-phenyl-ethylimidazole)[43].
- Co(II)-protoporphyrin-IX-diester (*7e*) with poly(4-vinylpyridine)[44].
- Metal chlorophyllins (*8*) (metal; Fe(II, III), Co(II, III), Ni(II), Cu(II), Zn(II) with poly(4-vinyl-pyridine) (partly quarternized)[45].
- Fe(II)-tetraphenylporphine (*9*) with poly[4-(1-methylimidazolyl)styrene][46], silicagel containing imidazole covalent over a propyl group[47],
- Metal octaethylporphines (*10*), metal Mg(II), Fe(II, III) with poly(1-vinylimidazole-co-styrene) (partly alkyl or aryl substituted at position 2 of the imidazole)[41, 48],
- Fe(II, III)-tetrakis[o-(alkylamido)phenyl]porphyrins (*11*) with poly(vinylimidazole-co-styrene)[49, 50] or triblock polymer like polyethylenglycol-poly(1-vinylimidazole-co-styrene)-polyethylenglycol containing thiophenol bridges between the blocks[49, 50].

(*8*)	M	R	R'
a	differ.	-H	-H
b	2H⁺	-CH₃	-CH₃
c	Mg²⁺	-CH₃	-phytyl

(*9*)	R
a	-H
b	-OH
c	-NH₂
d	-COOH
e	-COCl
f	-SO₂Cl

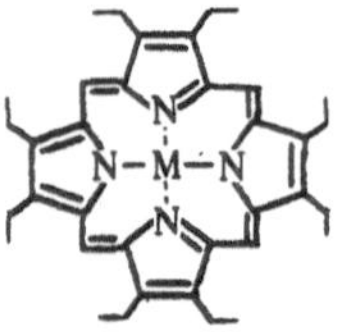

(10)

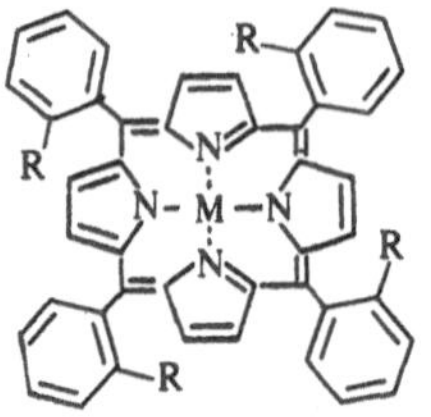

(11)   R = –NH–CO–R'

The *porphyrins with coordinative polymer bond* are prepared under inert gas as follows:

1. Organic solvent[19, 25)]

Fe(II)porphyrins
Fe(III)porphyrin
like (7b)
solved in DMF

Reduction by aqueous

————————→ Fe(II)porphyrin

solution of sodium hydrosulfite

————————————→

Excess of

polymer ligand like poly (4-vinylpyridine)

Formation of polymer ligand Fe(II) porphyrin in a mixture of DMF: CH_3OH: H_2O = 9 : 1 : 0.1 (Fe(II)prophyrin $\sim 10^{-5}$ mole/1, ligand $\sim 10^{-4}$ mole/1[25)])

With cobalt the Co(II)porphyrins were directly inserted in toluene together with the polymer ligand[44)]. The reduction of Fe(III)porphyrins was controlled by UV/VIS-spectra in the region of 410 nm (Soret band). By reducing with N-benzyl-1,4-dihydronicotinamide the reduction rate with poly(4-vinylpyridine) is surprisingly higher than with the low molecular ligand pyridine[30)]. Also degradation products of DMF like dimethylamine are able to reduce Fe(III)porphyrins[31)].

For a copolymer of 4-vinylpyridine and N-(p-vinylbenzyl)-3-carbamoyl-1,4-dihydropyridine containing both important functions (ligand and redox site) the rate of reduction is again increased[49)]. These reducing agents are important for the comparison with biological NAD(P)H.

2. Water (with water soluble polymer ligands like polylysine, polyvinylimidazole, partly quarternized polyvinylpyridine)[19, 20, 28, 34, 45, 29, 52)]

Polymer ligand in water
(perhaps with a small amount   ————————————→ polymer ligand-Fe(III) porphyrin mixtures
of DMF) pH 10–12   Fe(III)porphyrine
                            like (7b)

——————————→   Polymer ligand Fe(II)porphyrin complex in water at pH 10–12 (Fe(II)porphyrin
sodium $\sim 10^{-4}$ mol/1, polymer ligand $\sim 10^{-2}$ mol/1[34)]
dithionite

The natural Fe(III)hemoglobin(methylglobin) was used with synthetic polymers to get a polyion complex[53)]:

Solution of ferrihemoglobin   ————————————→
at pH 2.2 in water   K-poly(vinyl-
                        alcohol)sulfate

Separation of first stoichiometric polyion
complex (using basic groups of ferrihemoglobin)

solving at pH 12
$\longrightarrow$
Poly(diallyldimethyl-
ammoniumchloride)

Separation of second stoichimetric
polyion complex (using acidic groups of ferrihemoglobin)
The iron content of the complexes is $4.3 \cdot 10^{-5}$ mol/g. The complexes were used to investigate the binding of cyanide ion.

3. Solid[27, 39, 41, 32].

Polymer ligand (polyvinylpyridine poly-
vinylimidazole) and Fe(III)-porphyrins
in methanol or benzene

Aqueous solution of
$\longrightarrow$
sodium dithionite
or ascorbic acid

Formation of polymer ligand
Fe(II)porphyrin complex

Evaporation
$\xrightarrow{\text{of solvent}}$

Solid polymer ligand chelate (Fe(II)porphyrin
$\sim 10^{-4}$ mole per unit gramm of polymer matrix)

In another way polyanions like polystyrene sulfonic acid were added to an aqueous basic solution of the Fe(II) porphyrin and the polymer ligand and the resulting precipitate was isolated[27].

4. Binding to unsoluble ligands[39, 47].

Suspension of the
unsoluble ligand in
an org. solvent

$\longrightarrow$
Fe(II)porphyrin
(perhaps previous
reduction of Fe(III))

Unsoluble polymer ligand Fe(II)
porphyrin chelates

The metal porphyrin in the polymer ligand was used to investigate the modified situation in comparison with a low molecular ligand by
– spectroscopic methods like UV/VIS, ESR,
– determination of coordination number (formation of pyramidal or octahedral complexes (scheme 1, b→c→d),
– formation constants for the polymer complexes.
Details are reported by Tsuchida and Nishide[16].

The *coordination numbers* of the Fe(II), Fe(III) and Co(II)-porphyrins were determined by UV/VIS-spectra and ESR (Table 1)[16, 34, 37, 41, 44]. Very often the coordination

Table 1. UV/VIS- and ESR-data for Fe-porphyrin complexes containing imidazole ligands

Coordination number	Ligand	UV/VIS absorption λ_{max}(nm)	ESR signals (g values, DPPH as g-marker)
6	imidazole, poly(1-vinylimidazole) at pH 10 in water[34] or in DMF/MeOH[37]	426; 528; 558 (7a) 412; 530; 560 (7b)	2,71; 2,25; 1,60 (7b)
6	imidazole, poly(1-vinylimidazole-co-styrene) (2:1) in solid CHCl₃[41]		2,90; 2,27; 1,50 (10), Fe(III)
5	2-methylimidazole, poly(1-vinyl-2-methyl-imidazole) at pH 10 in water[34] or in DMF/MeOH[37]	431; 557 (7a) 399; 557 (7b)	
5	poly(1-vinylimidazole-co-styrene) (1:100) in solid CH₂Cl₂[41]		5,64 (10), Fe(III)

number for low molecular or polymer ligands are the same. In most cases only one base is coordinated. Using poly(L-lysine)[16] or poly(1-vinylimidazole)[34] with heme (7a) two units of these base are coordinated.

The coordination state of Co(II) protoporphyrin investigated by ESR changed from five to six by increasing the amount of poly(4-vinylpyridine-co-styrene) in solution[44]. ESR measurements state that the spin of Fe(III)porphyrin complexes with poly(1-vinyl-imidazole-co-styrene) depends on the amount of imidazole units in the copolymer (Table 1)[41]. Copolymers with more than 10% imidazole units show low spin adducts typical for six as coordination number. With less than 1% vinylimidazole in the copolymer high spin indicates five as coordination number. Using 1-vinylimidazole substituted in position 2 steric hindrance keeps the pyramidal coordination also at high concentration of the imidazole in the polymer (Table 1)[34, 37]. Five-coordinating complexes are active for O_2-uptake. So with artificial polymer ligands the situation may be similar to the coordinating property of globin.

Remarkable, the *formation constants* of porphyrin complexes with polymer ligands are 10^1 till 10^3 times higher than with low molecular ligands[16, 37]. This was explained by a higher concentration of the ligand now fixed in the polymer domain.

The **advantage** of coordinative bonds in metal porphyrins is the *easy preparation including high formation constants*. But the complex stability was not yet reported in the literature. It was observed that the complex of Fe(II)-protoporphyrindimethylester (7c) with poly(1-vinylimidazole-co-styrene) is easily separated from the starting material by reprecipating a benzene solution twice in methanol[41]. So a pocket like that in globin is not observed until now. In contrast with a covalent bond in metal porphyrins no separation can occur. Therefore the *application for coordinative bonds in metal porphyrins is limited* to unchanged solutions or solids.

Physical incorporation of different porphines (9) and also phthalocyanines (2) into polymer matrices were done by pearl polymerisations of glycidyl or 2-hydroxyethyl methacrylate and ethylene dimethacrylate[54, 55]. The amount of incorporated chelate is 0.05 to 0.4 wt %(50–70% of the inserted porphyrine). The surface hole is small enough to prevent diffusion of N_4-chelates from the interior of the matrix. Also polymers, such as polystyrene or polyvinylalcohole, were directly used to incorporate porphyrines or phthalocyanines.[56, 57].

2.1.2 Properties

Mainly the *binding of oxygen* (Scheme 1) was tested with the polymer ligand complexes. The important results are summarized as follows (s. also Table 5; comparision with covalent bonds in porphyrins. 3.1.3):

1) Fe(II)porphyrins in organic solvents
   - Fe(II)tetraphenylporphine (9a) connected with a crosslinked polystyrene containing imidazole in benzene is oxidized to μ-oxo-bridged dimer[46]. The nature of the dimerisation through comformational change of the polymer is not known.
   - In heme (7a) with poly(4-vinylpyridine) or poly(1-vinyl-2-methylimidazole) in DMF–CH$_3$OH–H$_2$O (9:1:traces) Fe(II) is partly oxidized after one cycle[25].
   - Oxygen is reversibly bound at Fe(II)tetra(pivalamidophenyl)-porphine (11) with poly(1-vinyl-imidazole-co-styrene) in toluene; the half life of the oxygenated polymer ligand complex is <1 day comparing with the low molecular ligand (1-ethylimidazole) complex being only 15 min[49, 50].

2) Fe(II) porphyrins in aqueous solution
   - Heme (*7a*) with poly(L-lysine) is slowly oxidized at pH 12 to the oxygenated complex; sigmoid curve of O_2 adsorption like in hemoglobin; also binds CO and CN^- [19, 20]:
   - Heme (*7a*) with partly quarternized poly(4-vinylpyridine) in H_2O/DMF = 9:1 has a much higher half life in the presence of an electrolyte than with pyridine complexes [28, 29]
   - Heme (*7a*) with a triblock copolymer containing styrene and 1-vinyl imidazole is reversibly, slowly oxidized [49].
   - Heme (*7a*) with poly(1-vinyl-2-methylimidazole); pH 10 in ethylene glycol: Comparing hexa coordinated heme with poly(1-vinyl-imidazole) to hexa or penta coordinated heme with low molecular imidazoles the above mentioned polymer ligand metal complex is absorbing O_2 reversible at 243 K. The UV/VIS spectrum is a good indicator for what is occuring (Fig. 1):

$$\text{deoxycomplex} \xrightarrow{\text{O}_2} \text{O}_2\text{-adduct} \xrightarrow{\text{CO}}$$

$$\text{CO adduct} \xrightarrow{\text{N}_2} \text{deoxycomplex}$$

The life time of the O_2 adduct is 12 min giving the oxidized complex. But this is the first success to have reversible oxygen binding in aqueous solution for some minutes. Also using coordinatively bound (*11*) at a copolymer of styrene and 1-vinylimidazole in H_2O saturated toluene $t_{1/2}$ is ~ 12 min.

3) Fe(II)prophyrins in solid state
   Fundamentally, on one side the rate of oxygenation is slower than in solution but on the other side the O_2 adducts are more stable against oxidation.
   - Heme (*7a*) with poly(4-vinylpyridine) (partly as polyion complex) [27] is reversibly oxygenated; after 4 cycles the reversibility is ~ 90%.
   - Fe(II)porphyrins with poly(1-vinylimidazole-co-styrene) [41] have a high spin Fe(II)porphyrins and are stable for weaks without oxidation.

4) Co(II)porphyrins
   Co(II)protoporphyrin-IX-dimethylester (*7e*) in poly(4-vinylpyridine-co-styrene) [44] in toluene at −90 °C forms an 1:1 O_2-adduct as observed by ESR and UV/VIS. At room temperature rapid oxydation to Co(III) in solution occurs.
   - Co(II)porphyrins with poly(1-vinylimidazole-co-styrene) [41] forms stable reversible O_2 adducts in the solid state also at room temperature.
   - Co(II)- and Fe(II)chlorophyllins (*8*) with partly quarternized poly(4-vinylpyridine) [45]: In aqueous solution at pH 10 the metal ions are easily oxidized. Fundamentally the rate of irreversible oxygenation is 10^{10} times slower for the polymer complex than with pyridine.

With spin labelled (*7b*) the metal chelate coordinated by polymer ligands in toluene or in the solid state shows lower ESR movement than the chelates coordinated by low molecular ligand [58]. So the possibility of dimer-dimer-interaction during oxygenation leading to irreversible oxidation is reduced. Aggregates of (*7a*)-pyridine complexes dis-

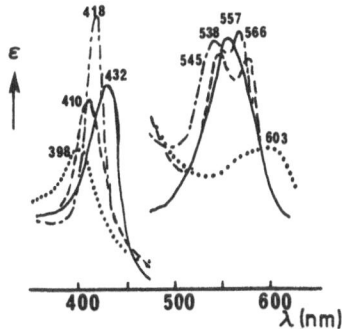

Fig. 1. VIS-spectra of heme (*7a*) ($8 \cdot 10^{-5}$ mol $\cdot$ l^{-1}) with poly(1-vinyl-2-methylimidazole) ($4 \cdot 10^{-2}$ mol $\cdot$ l^{-1}) in H_2O/ethylene glycol at pH 10 (303 K). (——) Deoxychelate; (– – –) O_2-adduct; (- - - -) CO-adduct; (···) oxidized product

sociate in aqueous solutions (pH 10) after addition of water soluble polymers[52]. Effective dissociation was observed in the presence of poly(4-vinylpyridine-co-vinylpyrrolidone). Therefore hydrophobicity of the polymers led to separation of porphyrins which is important to form a polymer pocket.

The before mentioned polyion complex between Fe(III)hemoglobin and K-poly(vinyl-alcohol)sulfate/poly(diallyldimethylammonium-chloride) is able to bind *cyanide ions* reversibly[53]. The cyanide ions are optimally bound at pH 8–9 till saturation at the Fe(III) occurs and are eluted with 0.1 N NaOH.

Remarkably, a synthetic polymer is able to stabilize porphyrins against oxidation. If Mg-octaethylporphin (*10*) in solid poly(1-vinylimidazole-co-styrene) with more than 20% imidazole units is irradiated in the presence of O_2 the π-cation radical of the prophyrin is formed quantitatively[48]. With less than 0.1% imidazole units the porphine is oxidized to formylbiliverdin. The results are compared with chlorophyll which is perhaps stabilized in the natural protein against O_2 in the same manner.

Hemin (*7b*), coordinatively bound to a macroporous resin containing imidazole units, shows at pH 9 in water *catalytic activities* for peroxidatiton[42]. Generally dimethylaniline is hydroxylated in a high yield in the following manner (Eq. 1):

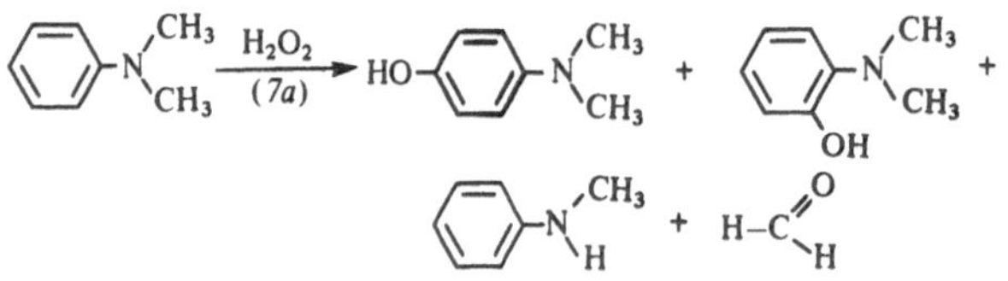

$$(1)$$

These orientating results are important because the type of reaction occurs also with pharmacologically active amines and is catalyzed in vivo by cytochromes P 450.

The before mentioned physically incorporated porphyrins or phthalocyanines in crosslinked polymers photosensitize efficiently the reduction of Fast Red A (FRA) (in the presence of L-ascorbic acid, AA, in DMF or water, Fig. 2)[54, 55]. Mainly the photosensitized reaction in water is interesting in the future water splitting. The reaction in water (Fig. 2) is explained by oxidative quenching (Eq. 2).

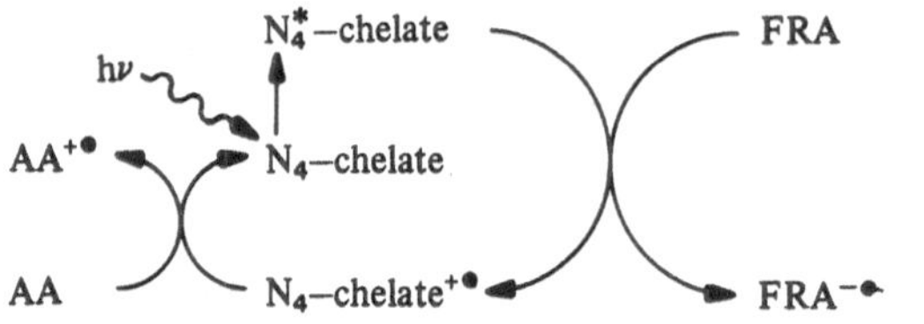

$$(2)$$

2.2 Phthalocyanines Pc (*2*)

The coordinative polymer binding of metal phtalocyanines was examined using polymer ligands like poly(ethylene imine)[59], poly(vinylamine)[59–61], poly(acrylamide) modified by dipropylenetriamine[59] and silicagel modified by γ-amino-propyl-triethoxysilan[59].

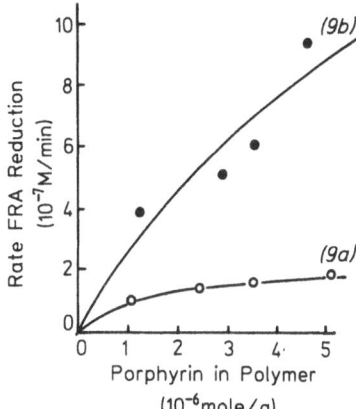

Fig. 2. Photoreduction of Fast Red A ($4 \cdot 10^{-5}$ M) by physical incorporated porphyrins (*9 a*, *b*) in a copolymer of glycidylmethacrylate, 2-hydroxyethylmethacrylate, ethylenedimethacrylate (0,15 g) and ascorbic acid ($2 \cdot 10^{-2}$ M) in 5 ml water (500 W xenon-lamp, 290 K)

Water soluble Pc are Co(II)tetrasulfophthalocyanine (*12 a*)[59, 61)] and Co(II)tetracarboxyphthalocyanine (*12 b*)[60)].

(*12*) R	
a	–SO$_3$H
b	–COOH
c	–NH$_2$
d	–COCl
e	–SO$_2$Cl

As with porphyrins, solutions or suspensions of reactands are mixed to get the complexes in an easy way. For CoPc(COOH)$_4$ (*12 b*) conclusive evidence of axial coordination was obtained from ESR showing 5-coordinate complex structure[60)].

Especially important is also the dimerization of CoPc like (*12 a*) (Eq. 3).

$$2 \, CoPc(SO_3Na)_4 \underset{PVA}{\overset{OH^{\ominus}}{\rightleftarrows}} [CoPc(SO_3Na)_4]_2 \qquad (3)$$

In alkaline medium the binuclear form is predominant. However, by increasing the concentrations of poly(vinyl-amine) the equilibrium is shifted to the monomeric form[61)]. Therefore a high concentration of polymer ligand is separating the Pc molecules in the polymeric coil (shielding effect). CoPc was tested as *catalyst* in the oxidation of thiols like 2-mercaptoethanol (Eq. 4; for conditions s. Fig. 3)[59–61)]. After coordination of the mercaptan anion (from amine and mercaptan) to the CoPc, formation of mononuclear oxygen adducts occurs in the presence of the polymer ligand. The activated oxygen may lead in a radicalic reaction to the end products.

$$\ldots-NH_2 + RSH \longrightarrow \ldots-\overset{\oplus}{N}H_3RS^{\ominus} \xrightarrow{CoPc}$$

$$\ldots-NH_3^{\oplus} \overset{R}{\underset{|}{S^{\ominus}}}{\longrightarrow} CoPc \xrightarrow{O_2} \ldots-\overset{\oplus}{N}H_3 \overset{R}{\underset{|}{S^{\ominus}}} \overset{\delta\oplus}{CoPc}-\overset{\delta\ominus}{O}-O$$

$$\qquad\qquad\qquad\qquad\qquad\qquad\qquad\qquad\qquad\qquad\qquad\qquad (4)$$

$$\longrightarrow \ldots-\overset{\oplus}{N}H_3 \overset{R}{\underset{|}{S^{\bullet}}}{\longrightarrow} CoPc-O-O^{\ominus} \xrightarrow{RSH}$$

$$\ldots-NH_2{\longrightarrow}CoPc + R-S-S-R + H_2O_2$$

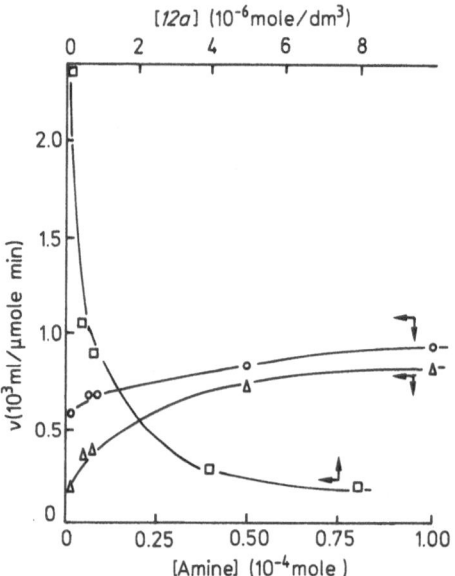

Fig. 3, a, b. Catalytic activity of poly(vinyl-amine)-CoPc (12 a) complexes for the oxidation of mercaptoethanol ($14,25 \cdot 10^{-3}$ mol in 11 ml H_2O) at 298 K. (a) as function of the poly(vinylamine) ($\bigcirc: \overline{P}_n = 50$; $\triangle: \overline{P}_n = 1680$) content with CoPc 10^{-8} mol, (b) as function of CoPc content ($\square$) with poly(vinylamine) ($\overline{P}_n = 570$) $4 \cdot 10^{-3}$ mol. $\overline{V}$ corresponds to O_2-consumption in ml/min per μmol of Co

Important results are:
- Increase of activity of the polymer system (12 a, poly(vinylamine), mercaptan) compared with the low molecular system (12 a), NaOH, mercaptan) by a factor of 50.
- Number of turnovers with the polymer system 10^6–10^7 comparing with the low molecular system 10^4–10^5.
- Increase of activity of the polymer system with increasing polyamine content, decreasing concentration of (12 a) and decreasing molecular weight of polyvinylamine (some results s. Fig. 3).

An interesting path is using metal phthalocyanines as models for hemoproteins. In this direction it was tried to get a coordinative bond between Fe- and CoPc(SO$_3$Na)$_4$ (12 a) at natural globin[62]. The interaction between globin or hemoglobin and the Pc led to green crystals after Sephadex gel separation. Spectroscopic methods shown that heme is partially displaced (Eq. 5) by MPc (disappearance of Soret band in VIS-spectra typical for porphyrins).

$$\text{(7a)-globin} \qquad \text{(12a)-globin}$$

$$\sim\!N\!\rightarrow\!FeP \quad + \quad \underset{(12a)}{FePc} \quad \longrightarrow \quad \sim\!N\!\rightarrow\!FePc \quad + \quad \underset{(7a)}{FeP} \tag{5}$$

Fundamental results of the reactions are:
- The Pc is bound as monomer from the solution containing dimers
- Like heme the MPc is located deeply inside the protein
- The FePc shows principally property of reversible O_2 *uptake* as seen by VIS-spectra and their O. D. values at 670–680 nm. Cyanide ions are bound irreversibly.

Physical incorporation of phthalocyanines and porphyrins in polymers was mentioned in Chap. 2.1.1 and 2.1.2[54–57]. Moreover, photovoltaic properties of Schottky bavier solar cells were checked by dispersing metal free Pc in a polymer binder[302]. At peak solar power (135 mW/cm^2) a power conversion efficiency of 1,2% has been obtained.

2.3 Schiff Base Chelates (6)

Tsuchida reported on synthesis and O_2 binding of N_2O_2-chelates of Co(salen) type[16, 49].

Like for porphyrin complexes the *preparation* of N_2O_2-chelates is rather easy: Mixing of polymer ligand like poly(4-vinylpyridine-co-styrene) with Co-chelates (6) in an organic solvent is sufficient. In contrast to porphyrins toluene or dichlormethane was taken instead of DMF because the more electrophilic Co in salenes interacts with DMF. In dichloromethane the entropic gain effect led to an increasing proportion of six as coordination number with increasing content of vinylpyridine in the copolymer[63].

The O_2 binding controlled by gas volumetric measurements, ESR-, and UV/VIS-spectra is characterized as follows:
- Formation of reversible 1 : 1 adduct (Scheme 1, e) at room temperature with high concentration of vinylpyridine comparing to 2 : 1 adducts (Scheme 1, f) with low molecular pyridine[49]. This is explained by the local base concentration near the chelate suppressing dimer formation
- Higher formation constants (k ~ 500 l/mol) of the polymer ligand comparing with low molecular pyridine (k ~ 13 l/mol) resulting from reduced motion of chelate through the macromolecular domain leading to a stronger intraction also with O_2 (Scheme 1: b → c → e).
- Comparing with Co-porphyrins better reversibility and stability of O_2 binding even at room temperature with higher formation constants and low ΔH and ΔS values.

2.4 Bis(1,2-Dioximes) (5)

The interesting properties of Co containing dioximes (5) are[12]:
- O_2-adducts of (5) containing Co(II),
- high nucleophilic reactivity of (5) containing Co(I) under formation of Co-C-bound with organic electrophiles,
- electrophilic reactivity of Cobaloxime containing Co(III) under formation of Co-C bound with organic nucleophiles,
- wide reactivity of organo-cobalt complexes.

When trying polymer binding of low molecular (5) two principles are realized:
- formation of Co-C bound to substituted polystyrene (s. 3.4.),
- binding at poly(4-vinylpyridine)[64].

Complexes of chloroaquocobaloxime/poly(4-vinylpyridine-co-styrene) (13) in DMF were studied to elucidate the effect of the polymer chain on complexation (Eq. 6)[64].

$$(5)\ (Co^{3+}Cl^-) \longrightarrow$$

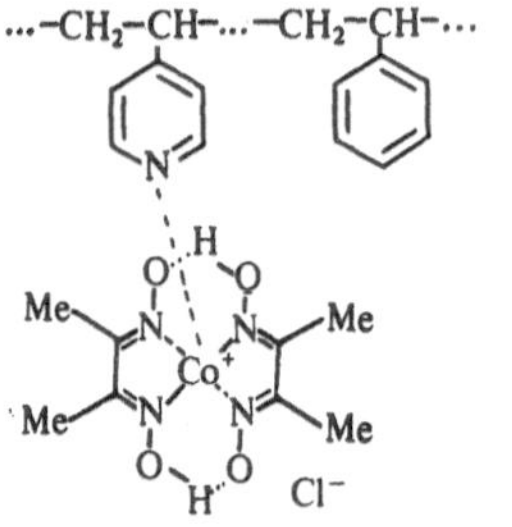

$$(6)$$

(13)

The equilibrium constants K (Table 2) were calculated from the change in absorption at 420 nm in electronic spectra[64]. The very low value of K in the homopolymer of 4-vinyl-pyridine (PVP) is explained by steric hindrance near the coordinated cobaloximes through the neighbouring base. If the content of 4-vinylpyridine in the copolymer is lower than 20%, the coordinated cobaloxime was easily removed because coordination occurs mainly at the surface of the polymer domain. In unpolar benzene instead of DMF, however, the conformational change of polymer chain brings the vinyl-pyridine units more to the interior of the polymer domain.

By using benzylbis(dimethylglyoximato)Co(III) and poly(4-vinylpyridine-co-styrene) in THF it was found that the higher complex formation constant is due to lower complex dissoziation rate constants comparing with low molecular pyridine[63].

Table 2. Equilibrium constants (K) of chlorocobaloxime ($\sim 6 \cdot 10^{-4}$ mol/1) in DMF with pyridine ligands L ([L]/[Co] = 0.02 − 2.00) at 293 K

4-vinylpyridine unit in copolymer styrene	$K \cdot 10^{-5}$ $(1 \cdot mol^{-1})$
9.8	0.45
13	1.0
24	1.2
32	1.4
41	1.3
53	1.5
100	0.085
pyridine	0.68

3 Covalent Polymer Bond in Metal Chelates

Generally, the covalent polymer binding of metal chelates leading to the structure element (E), (F) in contrast to coordinative binding have the *advantage of a strong fixation*. So they may be used in solution without migration of chelate from the polymer domain. The *disadvantage is the more difficult way of preparation*. So the use of the polymer metal chelates decides the direction of preparation.

The synthesis is realized in two principal ways:
- *covalent polymer binding of an chelate to a polymer with reactive groups* (Scheme 3):

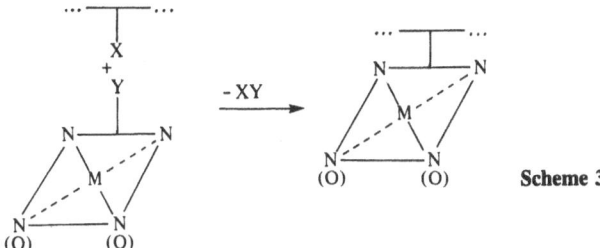

Scheme 3

- *Polyreaction of an reactive chelate monomer with other monomers to get a copolymer* (Scheme 4)

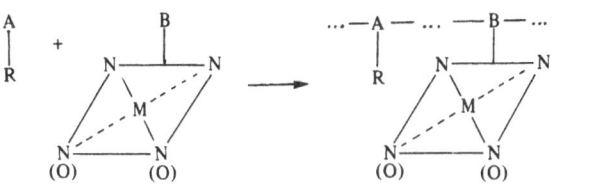

Scheme 4

3.1 Porphyrins

Because of the high importance of the pioneer work of Lautsch[65], few results are reported separate. Porphyrins are incorporated in a covalent polymer by:
1. Binding of porphyrin to a polymer chain
2. Copolymerization of the porphyrin containing vinyl groups
3. Polycondensation with porphyrins

Result to 1: Reaction of methylpheophorbide-a (*8b*) with poly(ethylenimine) leads under splitting of a 5-membered ring to the water soluble polymer chlorin derivative (*14*) (Eq. 7) containing one porphyrin per $10^2 - 10^4$ units of the polymer. Reaction with epichlorhydrin produces unsoluble materials with network structure. The reaction is controlled by electronic spectra (typical red shift from 632 to 644 nm while turning *8b* to *14*).

(*8b*) $\cdots$-CH$_2$-CH$_2$-NH-$\cdots$ $\longrightarrow$ (14) (7)

Polymers (*16*), highly stable against dilute acids and alkali, are produced by reacting mesoporphyrin-IX-diazid (*15 a*) with poly(ethylenimine) in dioxane (followed by the reaction with epichlorhydrin (Eq. 8).

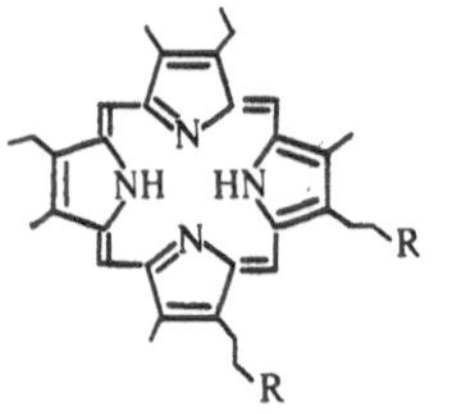

$$\tag{8}$$

(*15*)	M	R
a	$2 H^+$	$-CON_3$
b	$2 H^+$	$-COOH$
c	$Fe^{3+}Cl^-$	$-COOH$
d	$2 H^+$	$-N=C=O$
e	$Fe^{3+}Cl^-$	$-N=C=O$

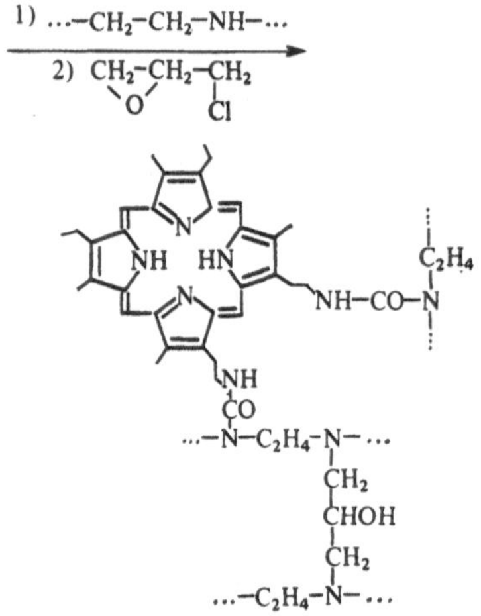

(*16*)

Result to 2: Thermal copolymerisation of styrene with the vinyl groups in methylpheophorbid-a (*8 b*) or protoporphyrin-IX-diesters (*7 f*) in a molar ratio of $10^3 - 10^5 : 1$ gives polymers (*17*) (*18*); they are soluble in toluene and stable in HCl.

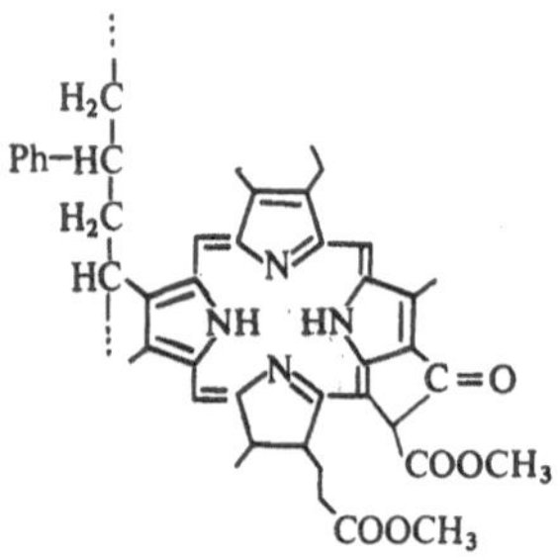

(*17*) 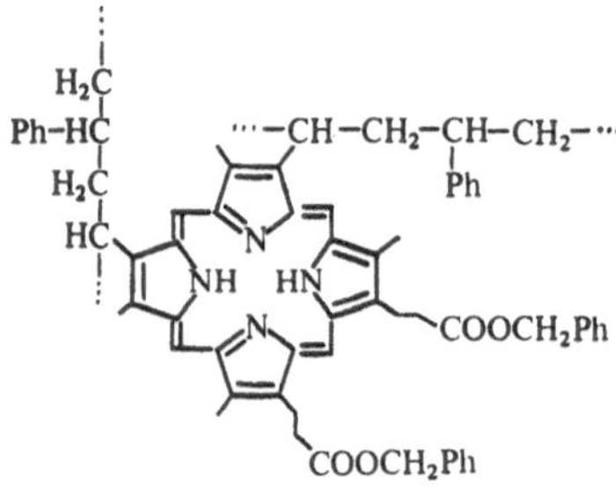   (*18*)

Results to 3: One method giving polyphenylalanines (*19*) starts from the diazide (*15 a*) and others reacting with 4-benzyl-2,5-dioxo-oxozolidine in pyridine (Eq. 9). Polypeptides containing the porphyrins in various ratios are obtained. Now from stereoisomeric N-carbonic acid anhydrides of phenylalanines the D-, L- and D,L-forms of the polyphenylalanines were prepared.

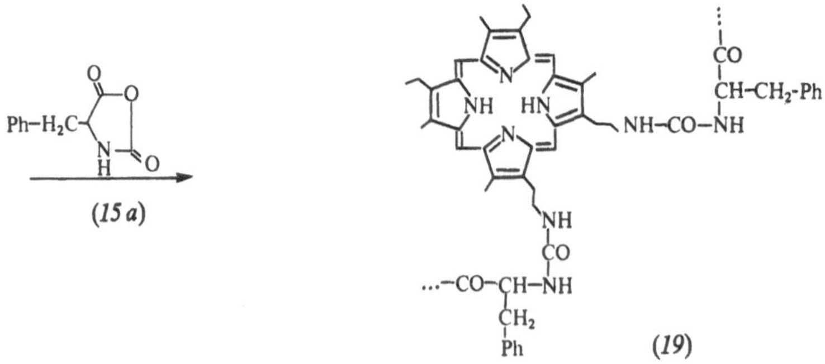

(*15 a*)

$$\tag{9}$$

(*19*)

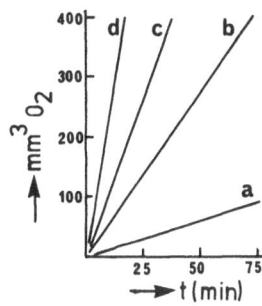

Fig. 4a–d. Oxidation of L-glutathion (50 mg) by mesohemin-polyphenylalanines (*19*) with Fe(III) in the porphyrin (with 10^{-6} g Fe(III) at 293 K in 7,3 ml buffer solution of ph $8^{65)}$). (**a**) blank value, (**b**) DL-form, (**c**) L-form, (**d**) D-form

After synthesis of the polymer metal free porphyrins metals like Fe(II), Co(II), Cu(II) were inserted in acetic acid or other solvents.

The polymers were mainly tested as models for enzymes. The relevant activity for the oxidation of L-glutathion (R–SH→R–S–S–R) was determined by measuring the O_2-consumption. As to be seen from Fig. 4 the optical pure polymers (*19* with Fe(III) in the porphyrin) show a higher activity than the DL-form. Generally the low molecular porphyrins have less than 50% activity compared with porphyrins polymer.

3.1.1 Covalent Binding of Porphyrins

The carboxylic groups of mesoporphyrin-IX (*15 b*) and mesohemin-IX (*15 c*) were converted into isocyanate groups (*15 d, e*)[66]. The reaction with hydroxyl group containing crosslinked polymers like poly(2-hydroxyethylmethacrylate) or sephadex (with hydroxypropyl groups) in pyridine at room temperature led to fixed porphyrins (*20 a, b*).

(*20*) M

a	$2H^+$
b	$Fe^{3+}Cl^-$

High molecular polyethylene and polypropylene grafted with active chloromethylstyrene (degree of grafting 15–30%) were converted with Co(II)-protoporphyrin-IX (*7g*) in DMF in the presence of triethylamine to give polymer binding through an ester linkage to (*21*) (Eq. 10)[67]. One quarter of the chloromethyl groups are reacting.

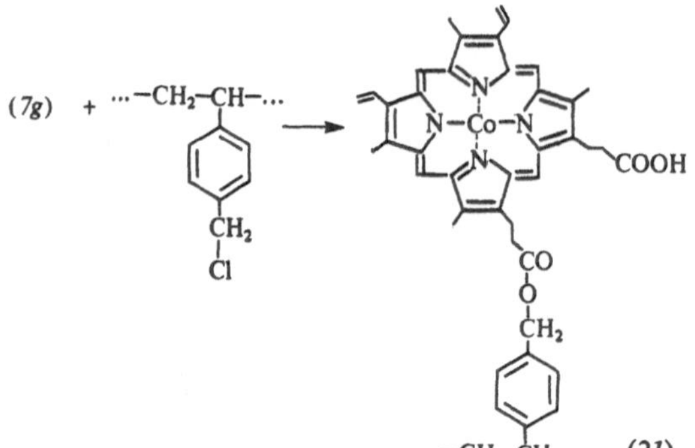

$(7g)$ + ····−CH₂−CH−···· $\longrightarrow$

(10)

$\cdots-CH_2-CH-\cdots$ (21)

Binding of mono- or binuclear porphyrins to modified polystyrene was tried (see also 3.1.2)[68, 69]. The aim to produce binuclear porphyrins is to simulate natural photoactive centers like chlorophyll a assumed to be an oligomeric active form.

An example for the fixation of mononuclear porphyrins is the reaction of metal free chlorin a (22) or protoporphyrin-IX (7h) with the chloromethyl groups of a watersoluble poly(chloromethylstyrene-co-1-vinylpyrrolidone) in THF in the presence of triethylamine (Eq. 11)[68]. The composition of the resulting polymer as (23) is only one possibility. Also Mg(II)-chlorophyll b was directly bound through its formyl group to the amino group of a copolymer containing 4-aminostyrene with formation of azomethine linkage[68].

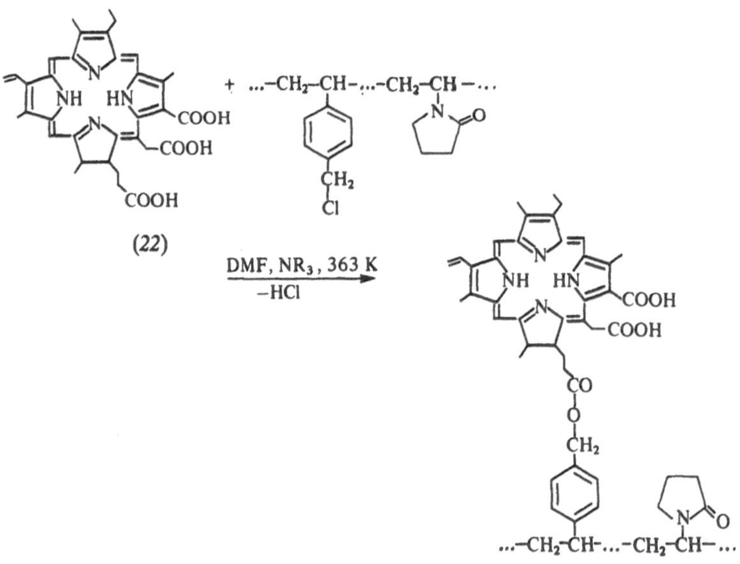

(11)

(23)

The dinuclear chlorophyll $(24a)$[69] was prepared from chlorophyll a $(8c)$ and 1,2-diaminoethane with subsequent hydrolysis of the phytol groups in a yield of 10%. Protoporphyrin-IX $(7h)$ gave the dimer $(25a)$ in 43% yield (Eq. 12). $(24a)$ and $(25a)$ were grafted on poly(chloromethylstyrene-1-vinyl-pyrrolidone) to give the dinuclear porphyrins (*26*), (*27*) in a content of 1–4%. Surprisingly no network formation of polyfunctional dimers occurred.

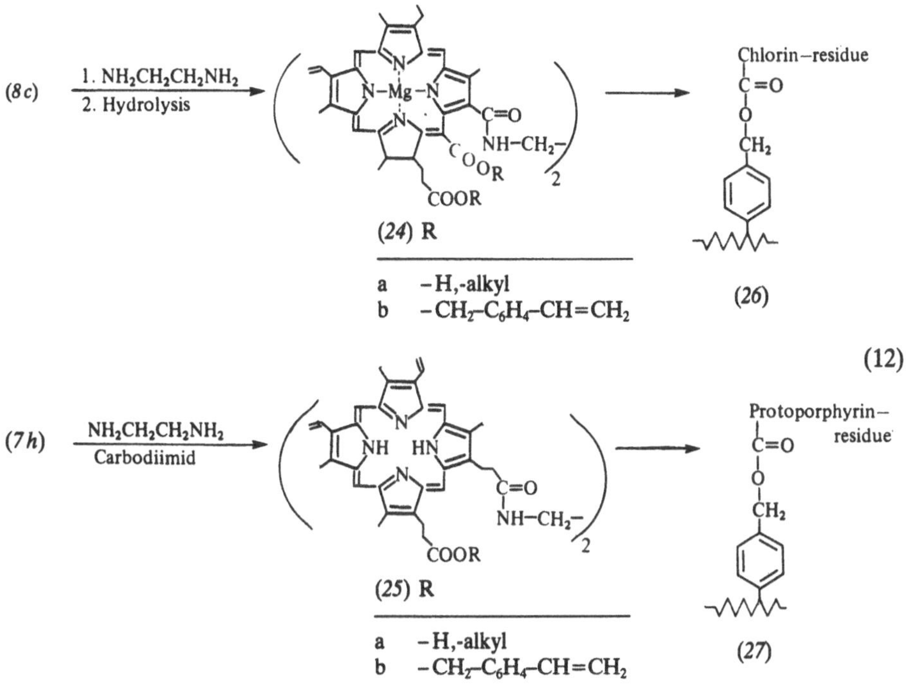
(8c)
1. $NH_2CH_2CH_2NH_2$
2. Hydrolysis
N-Mg-N
C=O
NH-CH$_2$-
COOR
COOR
2
(24) R
a -H,-alkyl
b -CH$_2$-C$_6$H$_4$-CH=CH$_2$
Chlorin–residue
C=O
O
CH$_2$
(26)
(7h)
$NH_2CH_2CH_2NH_2$
Carbodiimid
NH HN
C=O
NH-CH$_2$-
COOR
2
(25) R
a -H,-alkyl
b -CH$_2$-C$_6$H$_4$-CH=CH$_2$
Protoporphyrin–residue
C=O
O
CH$_2$
(27)

(12)

Porphyrin (*11*) was bound to the amino groups of a ternary block copolymer: polyethyleneglycol-poly(p-aminomethyl-styrene-co-styrene)-polyethylenglycol[50].

Heme (*7a*) containing 1-(3-aminopropyl)imidazole (as proximal base) and also histidine (as distal base) was converted with polylaminomethylstyrene to get polymer porphyrin (*28*) in a small degree of incorporation (< 0.1%)[70].

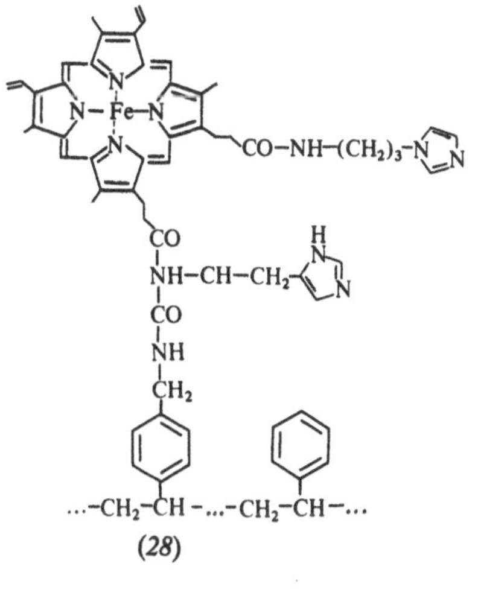
N-Fe-N
CO-NH-(CH$_2$)$_3$-N
CO
NH-CH-CH$_2$
CO
NH
CH$_2$
...-CH$_2$-CH-...-CH$_2$-CH-...

(*28*)

Also synthetic porphyrins were used for polymer attachment. The easiest way for synthetic porphyrins is the reaction between pyrrol and benzaldehyde to ms-tetra-phenyl-porphine (*9a*). Substituted porphines (*9c–e*) are prepared from substituted benzaldehydes. Porous crosslinked polystyrene resins were used in the reaction with (*9c–e*) as shown in Eq. 13 to get the polymers (*29*)–(*31*) containing 2–6% porphines in covalent bond[71].

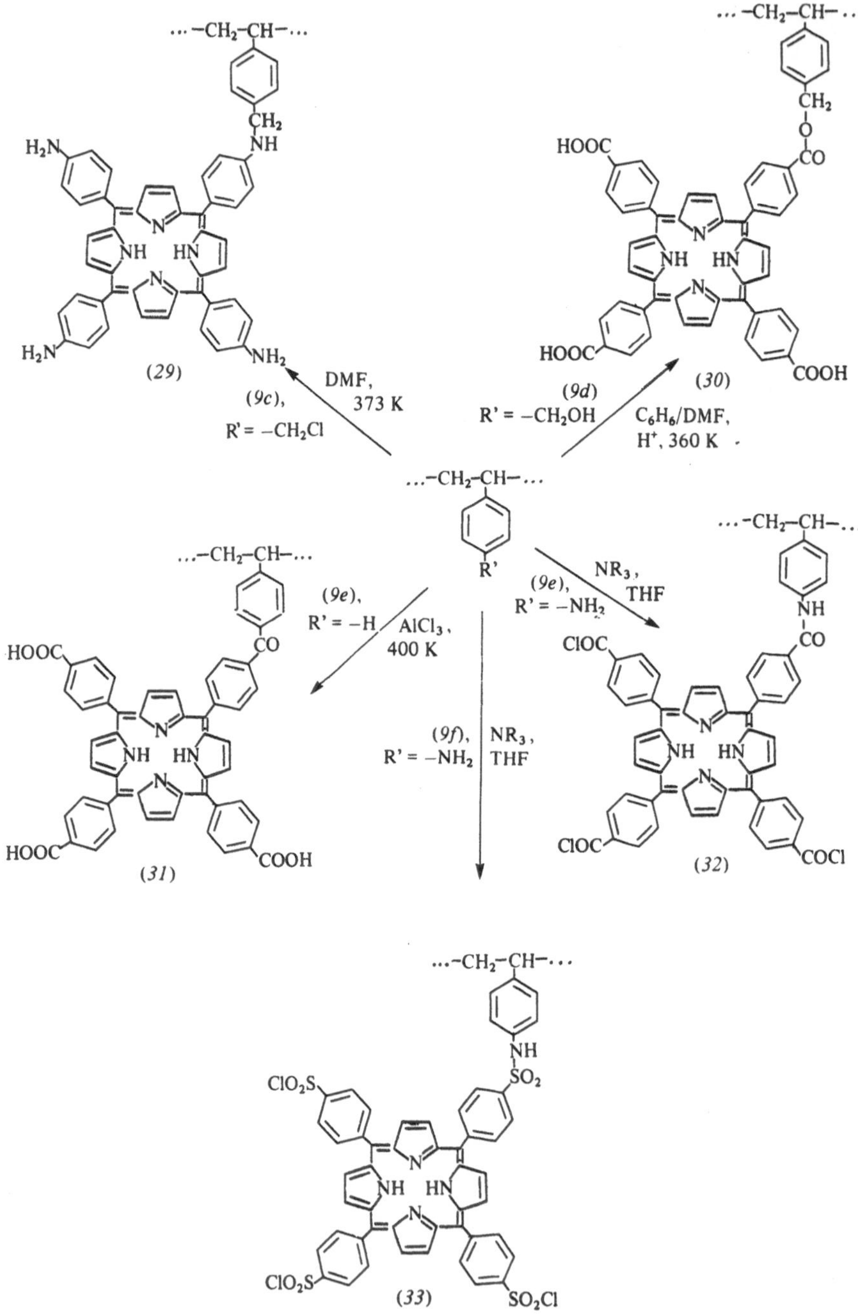

Afterwards metals like Co(II), Ni(II), Cu(II), Zn(II) were inserted. According to the ESR data there are no intermolecular interactions between the porphyrines[71].

Macroreticular polystyrene beads crosslinked with 20% divinylbenzene and containing p-aminostyrene units were converted with (9e, f) in THF into the polymers (32) and (33)[72] (Eq. 13). The unconverted acid chloride groups of the polymer bond porphyrins were converted with methanol into the corresponding esters.

An interesting new way was assumed by amide linkage between the Co-chelate of (9c) and (–CO–Cl)-group modified glassy carbon electrode[73]. In another case Fe-porphyrins bound at glassy C-electrodes were tested for O_2 reduction[74].

3.1.2 Polymerisation of Vinyl Group Containing Porphyrins

Two possibilities were realized using
– *natural vinylporphyrins*
– *synthetic vinylporphyrins*

Natural protoporphyrins contain one or two vinyl groups. Copolymerisation was first successfully performed by Lautsch[65]. More detailed investigations started recently[16, 41, 49, 50, 68, 69, 75–82]

Fe(III)-protoporphyrin-IX-dimethylester (7d) (0.03 mol%) was copolymerized with π-conjugated monomers like styrene or methyl methacrylate in bulk and acrylamide in methanol using radicalic initiators[75]. Increasing the ratio of [7 d]/[styrene] the molecular weight decreased (Table 3) indicating a chain transfer effect of the porphyrin (chain transfer coefficient $C_s = 2.3$). On the other side the content of covalent bond is increasing (Table 3). Therefore it is supposed that the observed reduction of Fe(III) to Fe(II) led to an addition of one porphyrin only at the chain end to give the polymer (34) (Eq. 14).

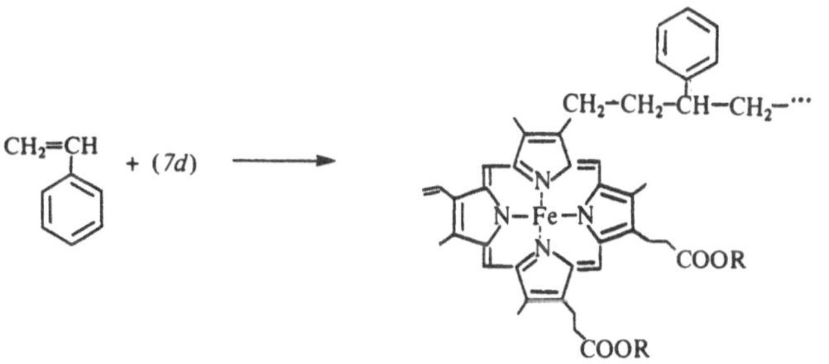

$$\text{(14)}$$

(34)

Copolymerisation with π-unconjugated monomers like 1-vinylpyrolidone, 1-vinyl-imidazole, 1-vinyl-2-methylimidazole was no success[75]. So terpolymerisation with additional styrene or acrylamide was carried out with 10^{-5}–10^{-6} mol/g porphyrin[41, 75]. In the electronic spectra the Soret band drifted from 393 nm (7d) to 387 nm due to the change at the periphery of the porphyrin.

The content of hemin (7b) in its copolymer with styrene mostly agrees with the monomers composition (Table 3)[77]. A reduced polymer yield and viscosity with increasing content of hemin in monomer mixtures was also explained by chain transfer.

Table 3. Radicalic copolymerisation of styrene with Fe(III)-protoporphyrin-IX-dimethylester $(7d)^{a\,75)}$ or hemin $(7b)^{b\,77)}$

Porphyrin Type	Porphyrin in monomer mixture (mol%)	Porphyrin in polymer (mol%)	Polymer yield (%)	$\overline{M}$ $\times 10^{-4}$	$\eta_{sp}/c\ (\times 10^2)$ (dl/g)
–	0	0		12.3	
$(7d)$	0.025	0.1		8.3	
$(7d)$	0.05	0.12		5.9	
$(7d)$	0.125	0.31		3.0	
$(7b)$	0.49	0.53	75.6		9.96
$(7b)$	1.95	1.94	60.7		8.66
$(7b)$	3.44	3.31	41.7		7.65

ª   Copolymerisation in bulk with AIBN at 333 K
ᵇ   Copolymerisation in pyridine with AIBN at 358 K

An interesting new way was shown by ^{60}Co γ-ray induced copolymerisation of hemin in the presence of another polymer: (all reactions controlled by UV/VIS-spectra)[76].

Aqueous solution of hemin at pH 10 $\xrightarrow{\text{Copolymer of}\ \text{1-vinyl-2-methylimidazole and 1-vinylpyrolidone}}$ Hemin in coordinative bond

$\xrightarrow[\text{Addition of CO}]{\text{Reduction of Fe(III) with hydrosulphite,}}$ Hem-CO-complex in coordinative bond $\xrightarrow[\text{irradiation by }^{60}\text{Co }\gamma\text{-ray}]{\text{HEMA or 1-vinylpyrrolidone}}$

covalent heme ($\sim 10^{-6}$ mol) in a transparent film (thickness ~ 2 mm) containing 70% solvent.

The copolymerisation of synthetic vinylpophyrins is mainly studied by Tsuchida and coworkers. Because oxygenation of Fe(II)-porphyrins in crosslinked polymer in suspension or in solids is much slower than with linear polymers in solution, much work was done to construct soluble polyvinylporphyrins with covalent bond.

The following monomers were prepared:

– vinyl derivatives of the synthetic tetraphenylporphine as mono(acrylamidophenyl)triphenylporphine (35);
tetra(methacrylamidophenyl)porphine (36);
tetra(p-styryl)porphine (37)[49, 78, 79];

Table 4. Copolymerisation of Vinylporphyrins with AIBN in DMF or benzene of 353 K[49, 78–80)]

Porphyrin	Inserted porphyrin content (mol%)	Comonomer	Porphyrin content in copolymer (mol%)
(35)	1.6	styrene	0.89
(35)	0.1	styrene	0.19
(35)	0.1	lauryl-methacrylate	0.13
(36)	0.07	styrene	0.1
(38)	0.03	styrene	0.045

– vinyl derivatives of natural porphyrins as protophyrin-IX-styrylamid $(38)^{49,\ 79-81)}$; metal free and metal containing (p-vinyl)benzylesters of chlorphyllins $(8\,a)$, chlorins (22) and protoporphyrin-IX $(7\,b,\ h)$ as $(39)^{68)}$ and dimers of chlorophyll $(8\,c)$ and protoporphyrin-IX $(7\,h)$ as $(24\,b)(25\,b)^{69)}$.

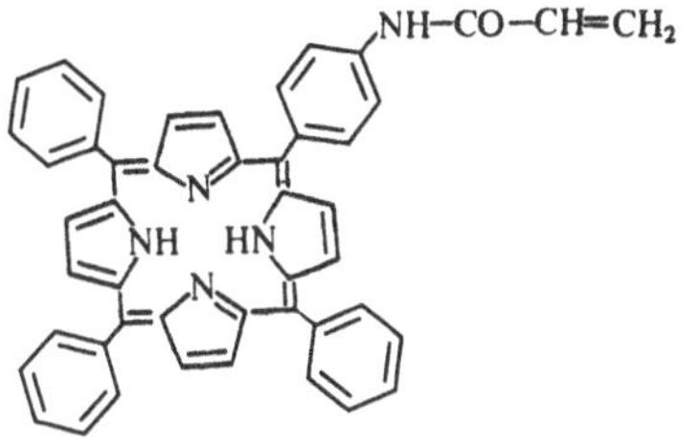

(35)

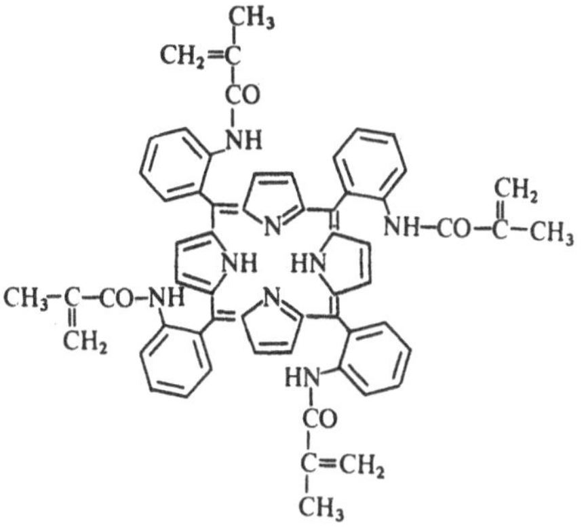

(36)

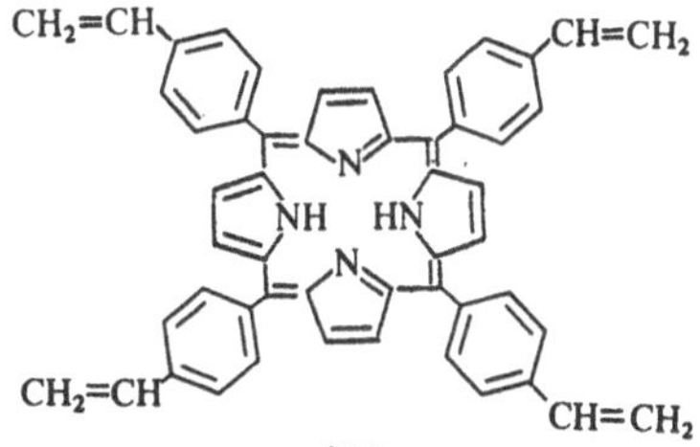

(37)

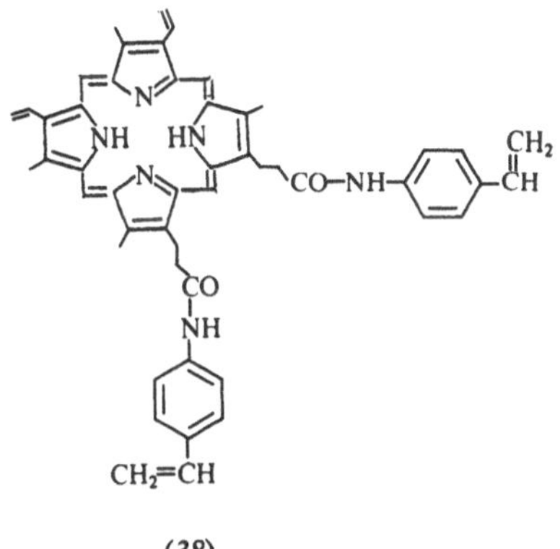

(38)

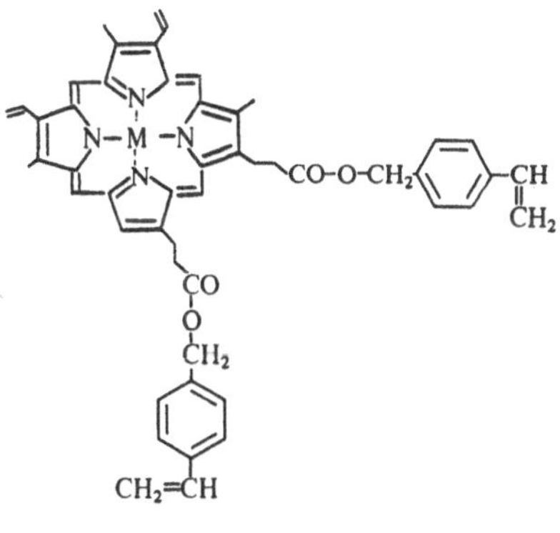

The copolymerisation of (35), (36) and (38) was carried out with styrene (also terpolymerisation with 1-vinylimidazole) and n-lauryl-methacrylate (Table 4):$^{78-80)}$

Mixing of co-monomers in benzene or DMF, adding AIBN	copolymerisation at 353 K for 2–4 h	With metanol separation, isolation of copolymer

gel permeatations chromatography →	Separation of high molecular copolymer from oligomerics and un- reacted vinylporphyrin	Incorporation of Fe(II), Co(II), Mn(II) in DMF under argon →	With water sep- aration, isolation of metal porphy- rin polymers; af- terwards purifica- tion

The yield of the metal free porphyrin polymers containing less than 1 mol% porphy-rin (Table 4) is 20–40%. With more than one vinyl group containing porphyrins (36)–(39) no network formation was observed.

The limiting factor for practical use may be the expensive synthesis and often low yield of vinylporphyrins. (35) was prepared in a three step synthesis from pyrrol and benzaldehyde/p-nitrobenzaldehyd in an overall yield of ~2%[78]. The picket fence por-phyrin (36) is obtained in a yield of only 5%[78]. (38) is synthesized from the expensive protoporphyrin-IX-dimethylester (7f) with 45% yield[80].

In the UV/VIS-Spectra the intensive red shift of the two absorptions of 590 and 650 nm accompanied with large change in the ratio of the absorption coefficients was assumed to be due to the perturbation of the polymer chain.

The (p-vinyl)benzylester (39) was synthesized in high yields from the corresponding natural porphyrin with 4-chloromethylstyrene; radicalic copolymerisation was tried with N-vinylpyrolidon (AIBN as initiator) and cationic copolymerisation with α-methylsty-rene (BF₃ as initiator)[68]. In most cases *metal free porphyrins* are best copolymerized with cationic initiators and *metal containing ones* best with radicalic initiators. This seems to be in contrast to former results. But it may be explained by the fact that vinylporphyrins are mostly unable to copolymerize with π-unconjugated monomers like vinylpyrrolidone. The vinylbenzylesters inserted in 5–10% were built in as comonomer nearly quantita-tively[68]. After conversion of the dimers (24a), (25a) with 4-chloromethylstyrene to the corresponding tetra(p-vinyl)benzylesters (24b), (25b) cationic copolymerisation with α-methylstyrene led to polymers containing ~2–4% of the porphyrin[69].

3.1.3 Properties

3.1.3.1 Binding of Small Molecules for the Transport of O₂ or CO

As mentioned earlier the reversible *oxygen binding* is of fundamental interest in order to construct an artificial oxygen carrier (see Chap. 2.1.2). Results on half life time of few covalent and some coordinative porphyrins are summarized in Table 5. Generally, the polymer Fe(II)-porphyrins are more stable against irreversible oxidation than the low molecular porphyrins. The slow oxidation of polymer Fe(II)-porphyrins is partly inhi-bited by steric hindering groups at the porphyrin (polymer from 36) (Scheme 1; e → f → g).

In aprotic solvents covalent bonds in porphyrins can be more stable against irrever-sible oxygenation than coordinative ones[41]. In aqueous solutions irreversible oxidation (reaction e → h in Scheme 1) is also a disturbing side reaction. Partly stable oxygen

Table 5. Half life time of oxygenated porpyrins at 298 K

Porphyrin	Metal	Type of binding	Conditions	Half-life	Ref.
TPP (9a)	Fe(II)	covalent	toluene + 1-ethylimidazole	30 sec	79
from (35) with styrene	Fe(II)	covalent	toluene + 1-ethylimidazole	5–7 min	79
from (36) with styrene	Fe(II)	covalent	toluene + 1-ethylimidazole	40 min	79
TPP (9a)	Co(II)	covalent	toluene + 1-ethylimidazole	–	79
from (36) with styrene	Co(II)	covalent	toluene + 1-ethylimidazole	1 day	79
tetrapivalamido-porphine (11)	Fe(II)	coord.	toluene + 1-ethylimidazole	15 min	49, 50
tetrapivalamido-porphine (11)	Fe(II)	coord.	toluene + copolymerstyrene vinylimidazole	< 1 day	49, 50
tetrapivalamido-porphine (11)	Fe(II)	coord.	H_2O satur. toluene	12 min	50
tetrapivalamido-porphine (11)	Fe(II)	covalent	3 v/v % dioxane (aq.)[a]	12 h	50
heme (7a)	Fe(II)	coord.	H_2O + pyridine	1 sec	28, 29
heme (7a)	Fe(II)	coord.	H_2O + polyvinylpyridine (partly quarternized)/NaCl	30 sec	28, 29
heme (7a)	Fe(II)	coord.	H_2O + polyvinylpyridine (partly quarternized)/ polystyrene sulfonic acid	2–3 h	28, 29
heme (7a)	Fe(II)	coord.	H_2O + poly(1-vinyl-2-methyl-imidazole)	12 min	34

[a] Covalent binding at the amino groups of a ternary block copolymer (polyethylenglycol-poly-(p-amino-methylstyrene-co-styrene)-polyethylenglycol) in hexadecakis(oxyethylene)dodecyl-ether

adducts were observed with hydrophobic and sterically hindering environment around the porphyrin[49, 50]. In another case addition of hydrophilic electrolytes like polystyrene sulfonic acid (as sodium salt), which may protect the porphyrin containing coil, leads to an increase of more stable oxygen adducts[28, 29]. Also immobilisation of covalent porphyrine (11) (at a ternary block copolymer) in micelles led to a drastic increase of O_2-adduct stability (Table 5)[50].

Covalent heme (28) containing covalent proximal base (N-bound imidazole) and distal base (histidine) is less stable against irreversible oxidation than, analogous polymers having only the proximal base; this is due to the proton donor property of the distal base accelerating oxidation[70]. A surprising result was shortly mentioned for a covalent heme (40)[83]. It contains amide groups bound to 3-(1-imidazolyl)propylamine and terminal histidine. The last one is bound to a polyethylenglycol-bis(glycinester) through an amide linkage. So proximal and distal histidine and also water solubility are imitated. Surprisingly some oxygenation and deoxygenation cycles with a sigmoide oxygen adsorp-

(40)

tion isotherme were observed. Evidently the results are contrary to the former mentioned ones.

Taking covalent or coordinative five-coordinated high spin Fe-porphyrins in the solid
state, oxidation is strongly inhibited but the times for oxygenation-deoxygenation cycles
are much longer than in solutions[41]. To avoid the disadvantage of slow gas adsorption in
dry solid films, the rapid CO uptake in water containing films of covalent heme in
another polymer matrix was successful carried out[76]. CO adsorption is also excellent in
aqueous solutions using terpolymers of styrene, vinylimidazoles and heme-diester[75]. But
in both cases no reversible O_2 binding was tested.

Polymer Co(II)-porphyrins in solution (polymer from 36) and in solid state[41] form
stable reversible oxygen adducts (Table 5). In contrast, low molecular Co(II)-porphyrins
are reversibly oxygenated only at low temperature.

3.1.3.2 Anion Binding for Cyanide Removal

Cyanide ion binding was done in order to examine Fe-porphyrin activity[66] and to use
resins as exchanger[77]. The absorption of the cyanide ion from aqueous solution increases
with increasing content of hemin (7b) in its copolymer with styrene[77]. A maximum value
of absorption is observed at pH 8. The ion-exchange capacity for a copolymer containing
2.5 mol% of hemin lies at 0.14 mg/g. Starting from a solution with 14 mg/l $CN^{\ominus}$ the
fractions after column experiments contain only 0.7 mg/l $CN^{\ominus}$ until the hemin is saturated. Recovery of the polymer is excellent with 0.5 N NaOH. Comparing the covalent
bond in (7b) with the before mentioned complex salt bond (2.1.2)[53] the first one has a
higher concentration of porphyrin per g so that it binds more cyanide ions.

3.1.3.3 Catalytic Activity for Oxidation and Energy Storage

Results on *catalytic activity* often are only fundamental, if properties of the porphyrin are
retained through covalent binding to polymers. Peroxidative activity with (20)[66] and
hydroquinone oxidation with (21)[67] was reported.

The polymers (29)–(31) were tested for the oxidation of butane thiol (2 R–SH + O_2
→ R–S–S–R + H_2O)[71]. At room temperature in hexane the Co(II)-containing polymers

(29) and (30) (as carboxylate anion) are active. But aging of the catalysts due to decomposition of active porphyrin under the condition of a radicalic oxydation was observed.

The polymer cobalt containing prophyrins (32), (33) were used in the photochemical energy storage system norbornadiene-quadricyclane (Eq. 15)[72]. This system for storage of solar energy is useful because of the high ΔH value between the two valence isomers on one side and the low price for the starting material (norbornadiene) on the other side.

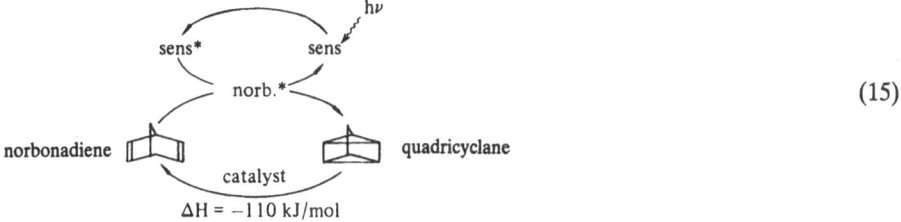

$$(15)$$

By the transformation of a sensitizer, light energy is stored by forming quadricyclane. In the back reaction using a catalyst the stored energy is set free[84]. The most active catalysts are low molecular on N_4- and N_2O_2-chelates mainly containing cobalt[84, 85]. Polymer metal chelates are easily separable from liquids when fixed on polymer beads.

The Co(II) containing polymer (32), (33) are highly active with turn-over numbers exceeding 10^4. That appears to be adequate for a first generation photochemical energy storage device. The addition of 0.1 g catalyst (containing 0.28–0.48% Co) takes 5 ml quadricylane to boiling after some seconds and 99% conversion occurs after 5 min. There is no great difference in activity between (32) and (33). Recycling studies reveal a deactivation pathway (Table 6). This corresponds to slow oxidation of Co(II) to Co(III) because the activity can be enhanced with strong reducing agents (Ti(III))[72]. Another investigation with low molecular Co-porphyrins showed that the Co(III) is more active than the Co(II)[86]. So the slow deactivation is still unclear.

3.1.3.4 Sensitizer for the Photoreduction

Mononuclear and dinuclear polymer porphyrins such as (23), (26), (27) and polymers from vinylderivatives (24 b), (25 b), (39) were tested as sensitizer in the photoreduction

Table 6. Recycling studies using polymer Co-porphyrin (33) for the conversion of quadricyclane at 303 K to norbornadien. 0.1 g catalyst (0,44 % Co) in 10 ml xylene containing 1 M quadricyclane. k: normalized value to 1 g catalyst to 1 l solution

Run no.	$k \times 10^4$ $(s^{-1} (g/l)^{-1})$
1	1.58
2	1.46
3	1.30
4	1.16
5	0.90

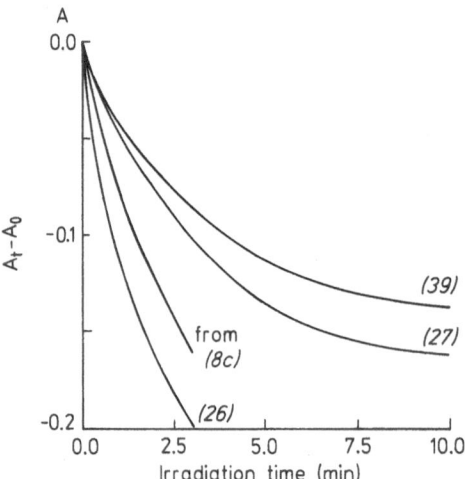

Fig. 5. Photoreduction of Fast Red A(FRA) in aqueous solution by the polymer mono-, dinuclear chlorophylls a from (*8c*), (*26*) and mono-, dinuclear protoporphyrine-IX (*39*), (*27*) in the presence of ascorbic acid (150 W xenon-lamp, 293 K)[68, 69]. Absorbance of FRA (at 520 nm with substruction of that at irradiation time A_0) plotted against irradiation time.
Porphyrin content in polymers: from (*8c*) 3.12%, (*26*) 3.94%; (*39*) 0.022%; (*27*) 0.024%. System containing ~1 g/l ascorbic acid; 0.04 g/l (FRA); 14 g/l polymer from (*8c*) or (*26*) and 38 g/l (*39*) or (*27*)

of Fast Red with L-ascorbic acid as reductive agent or hydroquinone in aqueous solutions[68, 69]. Figure 5 shows the results using Fast Red and the sensitizers (Eq. 2). With increasing irradiation time (150 W xenon lamp without filters) the absorption at 520 nm is decreasing. – The dinuclear porphyrins are more active than the mononuclear ones; this seems to be caused by the interaction excerted between two combined porphyrin rings. Cu(II) and Co(II) polymer porphyrins are not active. The concentrations are too high to show cyclic working of the sensitizer.

3.2 Phthalocyanines

In an early investigation poly(4-vinylphthalic acid anhydride) was converted with urea and CuCl to a polymer copper phthalocyanine[87]. This polymer decomposes at 583 K. Because of the unsolubility, it was only possible to detect many unreacted carboxyl groups in the polymer.

Another way was tried with condensation polymers (*41*) and (*42*) obtained from 2,5-diamino-3,4-dicyanthiophen and terephthalic acid dichloride or thiophen-2,5-dicarboxaldehyd[88]. Heating (*41*) and (*42*) with 1,2-dicyanobenzene and Cu(II)-salts at 473–523 K the corresponding phthalocyanines (*43*), (*44*) were obtained (Eq. 16). According to IR, elementary analysis and hydrolysis every second unit contains a phthalocyanine system.

Some authors describe the covalent polymer binding of phthalocyanines[60, 89–93]. Co(II)-tetraaminophthalocyanine (*12c*) was coupled with cyanuric chloride to the amino groups of crosslinked aminated polystyrene or polyacrylamide (with aniline substituted groups) to get (*45*) and (*46*)[89].

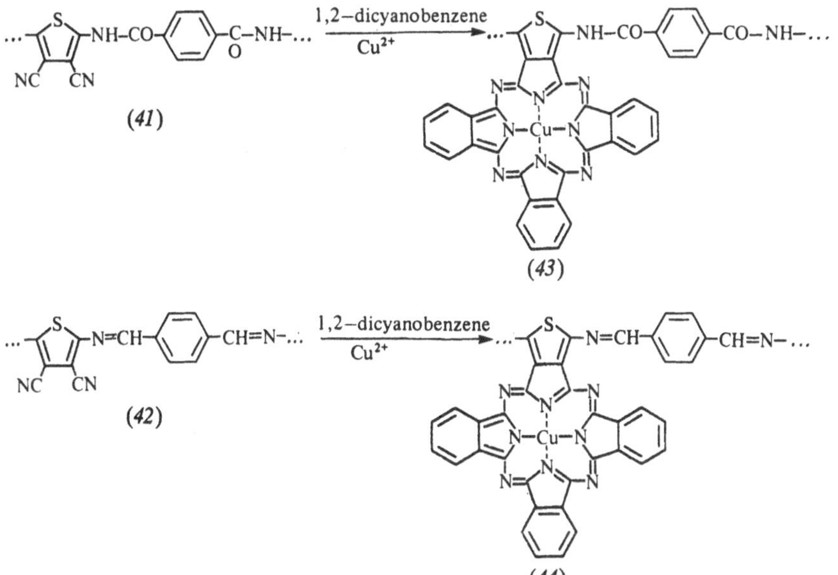

(41)

(43)   (16)

(42)

(44)

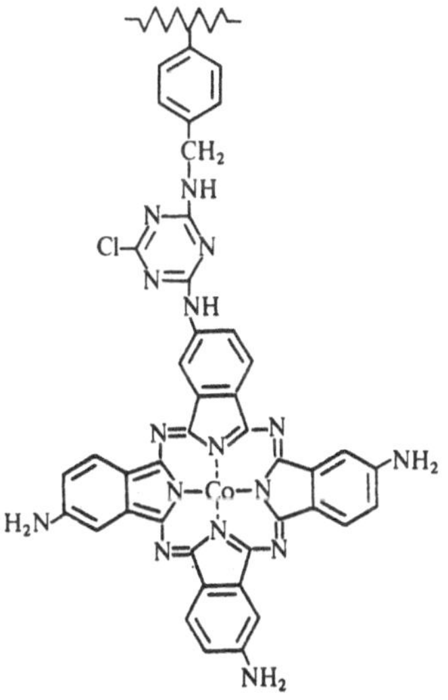

(45)

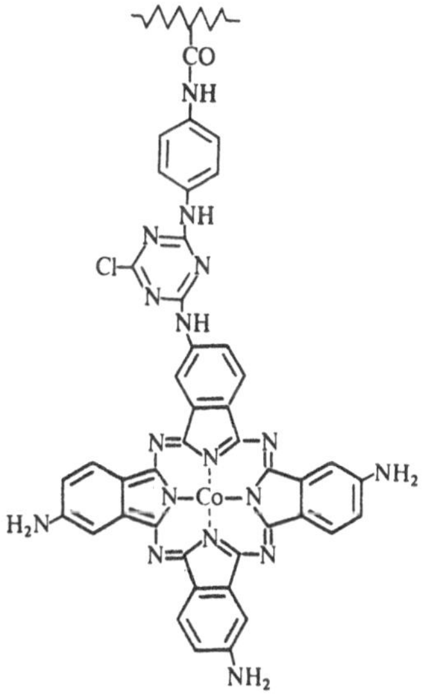

(46)

(45) and (46) were tested as *catalysts* for the oxidation of 2-mercaptoethanol to the disulfide in aqueous alkaline solutions as described in 2.2. Remarkably the catalysts are not deactivated after several runs. The polymer (46) shows a higher activity than the low molecular Co-phthalocyanine due to a high concentration of the mononuclear superoxo-complex; it is more active than a dinuclear peroxo complex (Eq. 17).

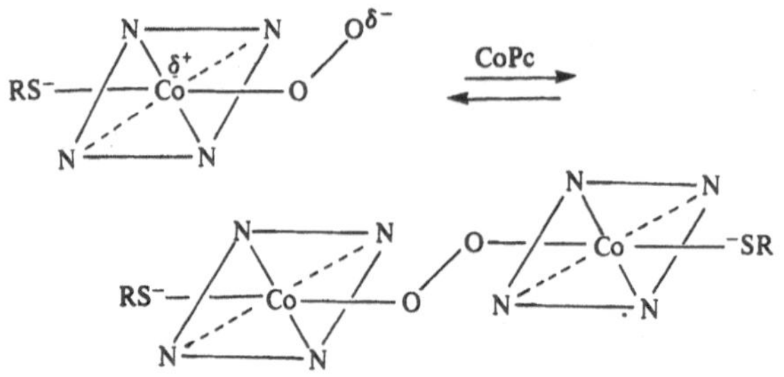

$$(17)$$

The ESR signal intensity shows that 30% of (46) exists in water in the mononuclear form. In DMF over night the intensity decreased by a factor of 5[90]. The increasing signal at $g \sim 2003$ corresponds to the oxidized phthalocyanine $(Co(III)Pc^{2+} \cdot O_2^{-})$. Consequently, the polymer is reducing the dinuclear formation drastically.

Co(II)-tetracarboxyphthalocyanine (12b) was condensed with poly(vinylamine) by dicyclohexylcarbodiimide in THF to give the polymer (47) containing $\sim 0.013\%$ Co (28% of the inserted phthalocyanine)[60]. Comparing the activity of (47) for mercaptoethanol oxidation with (12b) in coordinative bond with poly(vinylamine) (s. 2.2), no great difference was found. The advantage of (47) is the long time stability after adding some NaOH in contrast to Pc with coordinative bond (Table 7).

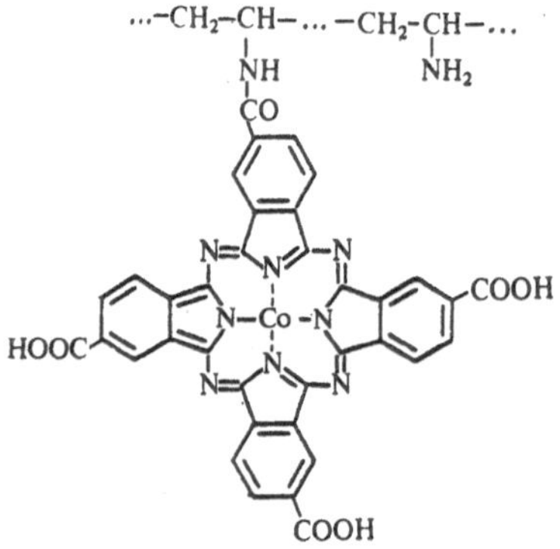

(47)

Table 7. Relative activity of (47) in the oxidation of mercaptoethanol (aqueous solutions with small amount NaOH) at 298 K

Run	Activity
1	1
2	0.65
3	0.48
4	0.42
5	0.46
6	0.42

The tetracarboxychloride (*12 d*) of the phthalocyanine (containing Fe(III) and Co(II)) was attached to linear polystyrene in nitrobenzene with AlCl$_3$ to give the polymer (*48*)[91]. Reaction mixtures having less than 4 mole% of (*12 d*) are not crosslinked by reaction of more than one acid chloride group. The UV/VIS spectra of these DMF soluble polymers allow to recognize the aggregation of phthalocyanines to dimers being inhibited due to steric hindrance of the polymer chain (s. 2.2).

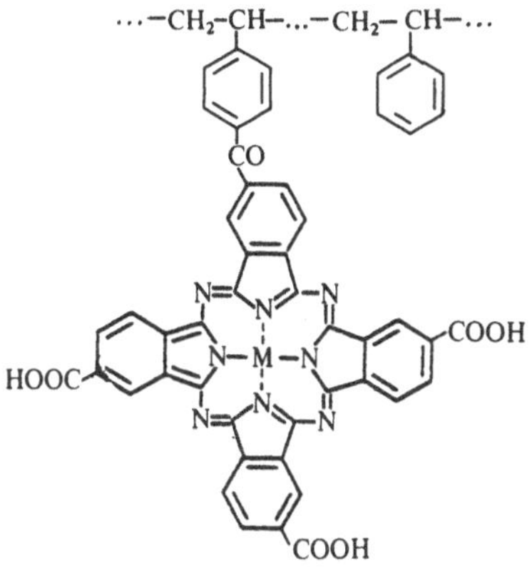

(*48*)

The catalase like activity was tested with (*48*) as catalyst[91]. From the results it is evident that the polymer bond in phthalocyanine led to a lower activation energy due to higher concentration of not aggregated active centers than with low molecular phthalocyanines. Continuous flow experiments in a column show that (*48*) keeps 60% of its original activity. The polymer is more stable than the low molecular phthalocyanine.

Macroporous highly crosslinked styrene-divinylbenzene copolymers grafted with poly(vinylamine) (Eq. 18) were synthesized in order to get unsoluble active catalysts of type (*47*) from tetracarboxy-phthalocyanine (*12 b*)[92]. In general, there is no great difference of activity between covalent and coordinative bonds in Pc for the oxidation of mercaptoethanol but higher activity compared with the low molecular system (*12 a*) in NaOH (s. 2.2). High surface area of the resin increased while low internal surface decreased activity. High concentration of poly(vinylamine) enhanced activity.

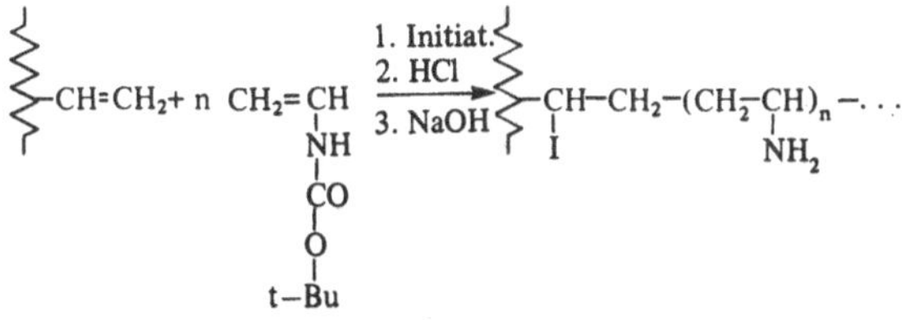

(18)

Attachment of chlorosulfonated MPc by sulfonation and sulfonamide linkage to macroreticular polystyrenes led to a polymer containing 0.0027 to 0.035% (*49*), (*50*) (Eq. 19)[93]. Here, the MPc are mainly aggregated at the edges of the beads. The activity of the polymers for the oxidation of cyclohexene at 351 K compared with low molecular

MPc is only higher after grinding the polymers; this is due to a better diffusion of the reactants.

In general, MPc are more stable against irreversible oxidation than porphyrins (Scheme 1, e → f → g, e → h). But the O_2-binding ability is smaller.

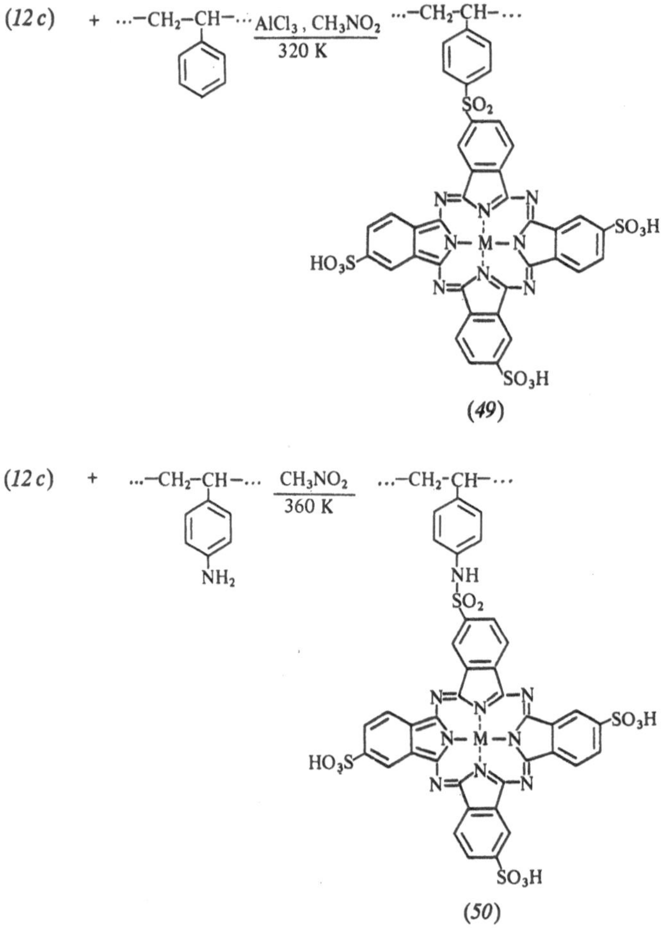

$$(49)$$

$$(19)$$

$$(50)$$

3.3 Schiff Base Chelates

The first production of polymer N_2O_2-chelates starts from a macroporous glycidyl methacrylate-ethylene dimethycrylate copolymer[94, 95]. Nucleophilic addition of ethylendiamine led to polymer amine, which is converted with salicylaldehyde to the Schiff base ligand (51) and then with Cu^{2+} and Co^{2+} to chelates (52) (Eq. 20).

By chelate formation, two, and by oxygen binding together, four ethylendiamine containing units of the polymer must be arranged around the reaction center.

Last time syntheses of polymer bond Schiff Base chelates were reported[96–101]. The main investigation was done by Wöhrle and coworkers.

```
...-CH₂-CH-...                ...-CH₂-CH-...              H
         |                             |                  \
         CO      NH₂CH₂CH₂NH₂          CO                  C=O
         |      ——————————————→       |                  /‖
         O                            O               [benzene ring]
         |                            |                  \
         CH₂                          CH₂                 OH     ——————————→
         |                            |
         CH   O                       CHOH
         |   /                        |
         CH₂'                         CH₂
                                      |
                                      NH
                                      |
                                      CH₂
                                      |
                                      CH₂
                                      |
                                      NH₂
```

$$(20)$$

```
...-CH₂-CH-...                ...-CH₂-CH-...
         |                             |
         CO                            CO
         |                             |
         O                             O
         |                             |
         CH₂          Co²⁺             CH₂
         |        ——————————→          |
         CHOH                          CHOH
         |                             |
         CH₂                           CH₂
         |                             |
         NH                            NH
         |                             |
         CH₂                           CH₂
    H    |                        H    |      H
     \   CH₂                       \   CH₂    |
      C=N                           C=N    N=C
   [ring]                        [ring]  \  /  [ring]
        OH                           O—Co—O

       (51)                            (52)
```

The preparation of polymer bound Ni-chelates (54) of the acacen type started from chloromethylated polystyrene and pentane-2,4-dione leading with catalytic amounts of sodium ethoxide to the polymer diketone (53) (Eq. 21)[301]. Seven % of pendent diketone groups of soluble linear (53) were converted with 1,2-diaminoethane and Ni(II) to the polymer Schiff base chelate (54). In the reaction of (53) to (54) two pendent diketones must react. But surprisingly no network formation was reported.

```
...-CH₂-CH-...                        ...-CH₂-CH-...
       |                                     |
   [benzene ring]                        [benzene ring]
       |           1. NH₂CH₂CH₂NH₂           |
       CH₂        ————————————————→          CH₂
       |              2. Ni²⁺                 |
       C                                      C
      / \                                  O   N
     O   O                                  \ /
                                            Ni
      (53)                                  / \
                                         O   N
                                         |    |
                                        CH₂  [ethylene bridge]
                                         |
                                    [benzene ring]
                                         |
                                    -CH₂-CH-

                                        (54)
```

$$(21)$$

Starting from chloromethylated polystyrenes and malonitrile or 3,3'-iminodip-ropionitrile after introduction of different metal ions, the polychelates (55) and (56) were obtained[96, 97]. The main problem is the long reaction path (four step reaction) and the difficulty to varify the type of the bridge between the original salicylaldehyde groups.

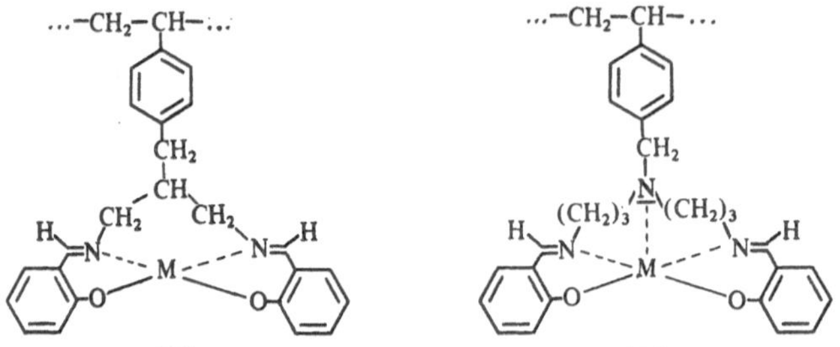

(55)   (56)

The Co chelate (56) was used as a catalyst for the oxidation of 2,6-dimethylphenol to the corresponding quinone (BQ) and diphenoquinone (DQ) in benzene at 298 K[97]. Phenol oxidation is well known with low molecular Schiff Base chelates[102].

Comparing the used gel resin (56) with low molecular N_2O_2-chelates it shows lower activity but much higher selectivity by changing the concentration of the chelate centres in the catalyst: BQ/DQ ~ 16 by a 0,29 mmol containing Co(II)/g resin, BQ/DQ ~ 0,63 by a 0,14 mmol containing Co(II)/g resin.

As expected for the high spin Co(II)-chelate (56) no ESR signal was observed (at < 77 K). But by exposing (56) to O_2 at low temperatures g values and hyperfine structure are due to known oxygen adducts of low molecular chelates. The Fe(II) chelate (56) is easily oxidized to Fe(III).

Polychelates containing the structure elements (57)–(60) were described[98–101]. The four methods have the advantage that many structural parameters, such as metal ion, salicylaldehyde, bridge (diamine) are easily variable. In order to get (57), linear and macroporous polystyrenes were acylated with 5-chloromethylsalicylaldehyde (Eq. 22)[98]. The reaction (22) with ethylendiamine, salicylaldehyde and Co^{2+} led to (57) containing 3.1–4.5% Co (every tenth to fifteenth benzene ring contains Co in a covalent bond (salen)). All reaction steps are quantitative. The main problem is crosslinking during the alkylation.

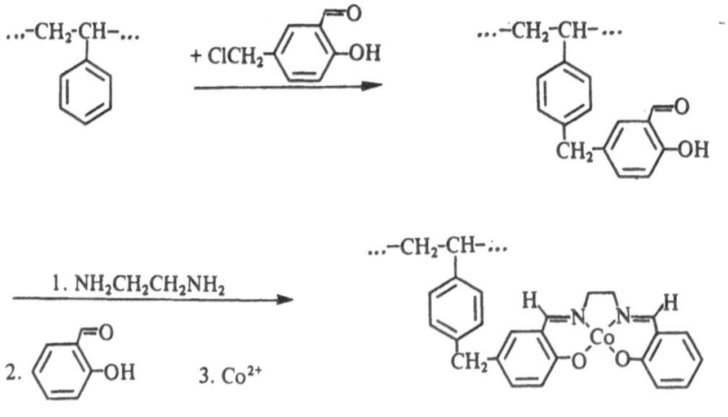

(22)

(57)

For the synthesis of (58) the 2-butyliminomethyl-4-vinylphenol was copolymerized with styrene in bulk (Eq. 23)[99]. The copolymerisation parameters starting from a ratio styrene/vinylphenol = 5 to 100 are $r_1 = 1.25$, $r_2 = 0.19$. Therefore by using an excess of styrene the copolymerisation goes to high yield without formating homopolymer blocks. The azomethine group in the copolymers was saponified giving the poly(vinylsalicylaldehyde-co-styrene). The following reaction with ethylendiamine leads to the polymer ligand under network formation. Afterwards Co^{2+} was introduced. The main disadvantage for (58) is the long reaction path.

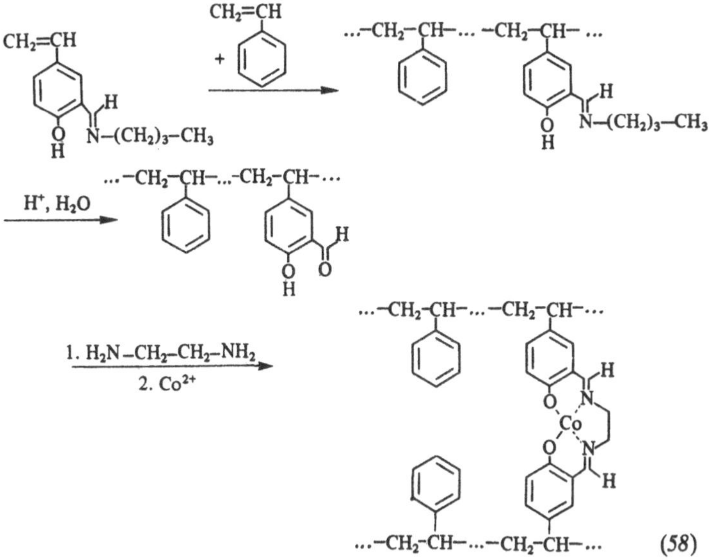

(23)

(58)

Various 4,4-divinylsalenes were used as starting material for (59) (Eq. 24)[99]. The advantage is a one step reaction in the copolymerisation with styrene leading directly to the ligand. With ratios of styrene/divinylsalenes = 5–100 the copolymerisation parameters were found $r_1 = 1.61$ and $r_2 = 1.68$.

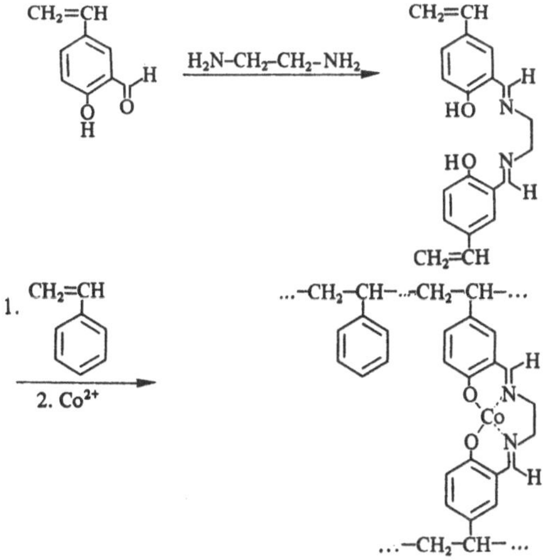

(24)

(59)

Polymer Schiff Base ligands containing a nitrogen atom in the bridge are easily synthesized by N-alkylation from various types of linear or crosslined chloromethylated polystyrenes and low molecular N_3O_2-ligands[100]. Afterwards Co(II) was introduced to obtain the chelates with 2–4% Co (Eq. 25) and (60) is isolated.

The polychelates (57)–(60) were examined for oxygen binding and activation[101].

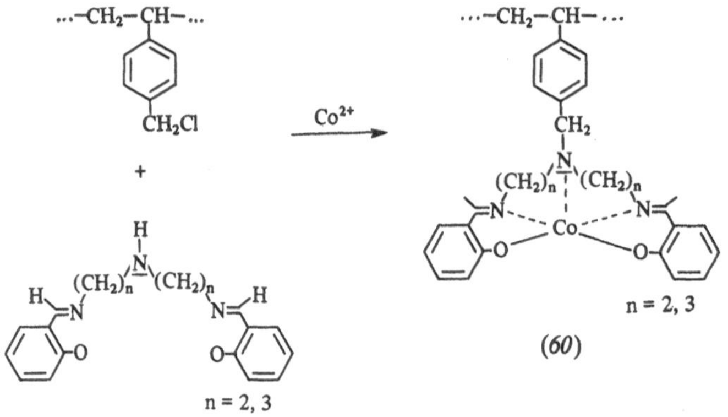

(25)

n = 2, 3

The polymer binding of low molecular N_2O_2-chelates like Co(salen) (6) starts from its monoanion and poly(chloromethylstyrene) in THF at 193 K. The high reactivity of the reduced (6) leads to the polymer (61) by formation of a Co–C bond in 100% yield. But no use of such an easily accessible polymer is yet known.

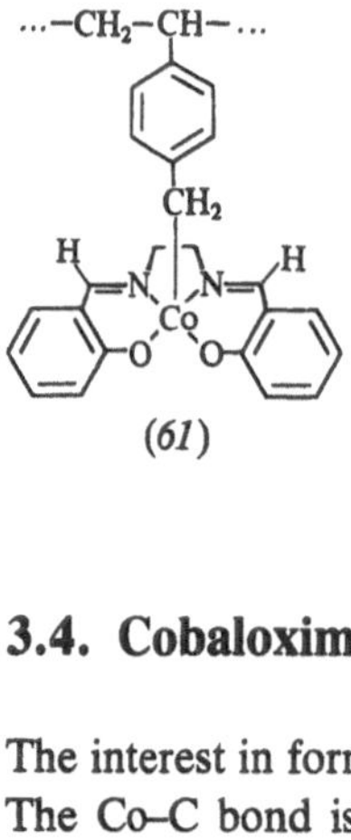

(61)

3.4. Cobaloximes

The interest in formation of Co–C bond to a polymer is related to coenzym B_{12} (s. 2.4). The Co–C bond is highly reactive. Such a polymeric carrier with Co–C bond (62) is synthesized by reacting poly(chloromethylstyrene-co-styrene) with Cobaloxime (containing Co(II)) in benzene pyridine[103].

By irradiation with a tungsten lamp the homolytic cleavage of the Co–C bond was observed in electronic spectra. In benzene the polymer domain is stabilizing the Co–C bond. The photodecomposition rates of the polymer (62) ($k_{dec} = 10^2$ sec^{-1}) are smaller than that of low molecular benzyl cobaloxime ($k_{dec} \sim 5 \cdot 10^{-2}$ s^{-1}). However, in DMF no difference is observed. The stabilisation in benzene is explained by location of the cobaloxime centers in the inner sphere of the polymer domain surpressing the reverse reaction.

...−CH₂−CH−...

(62)

4 Polymer N₄-Chelates with the Metal in the Main Chain

Low molecular N₄-chelates may react to polymers where the metal atom is a part of a polymer chain. Generally, *three ways for construction* of such a polymer are possible (Scheme 5):

a) The low molecular N₄-chelates contain metals with a higher oxidation state than three, such as the IVa-group elements in the polymers (63)–(66).

b) With three valent metal atoms polymers having a combined covalent-coordinative bond in the main chain are obtained.

c) Bivalent metal atoms may react with bivalent donor bases to polymers.

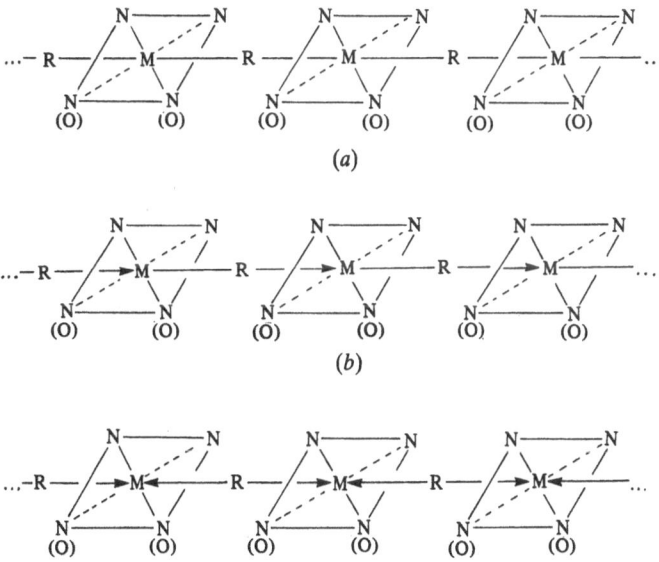

Scheme 5

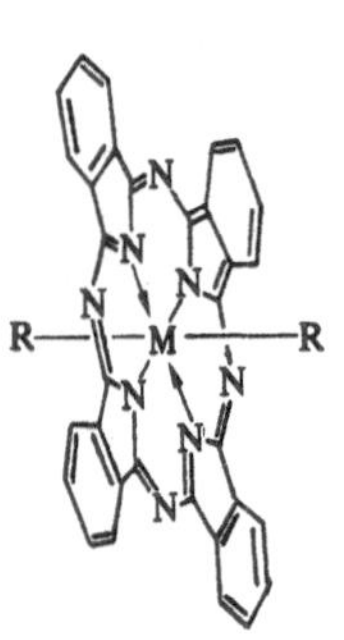

(63)   (64)

Pc   M = Si, Ge, Sn   Tpp   M = Ge, Sn
       R = –OH, –Cl   R = –OH, –Cl

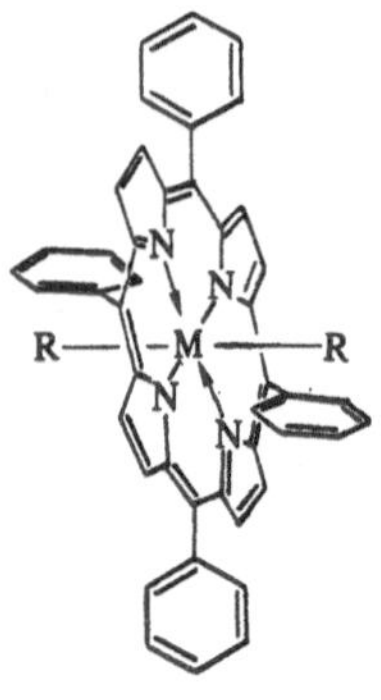

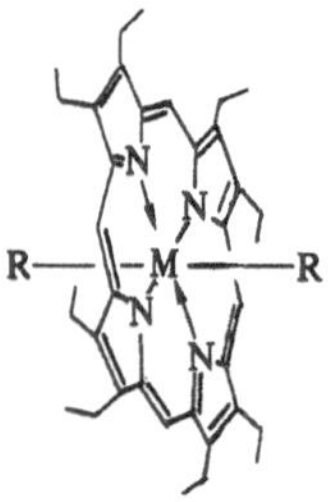

(65)   (66)

Etiop M = Ge, Sn   Hp   M = Ge, Sn
       R = –OH, –Cl   R = –OH, –Cl

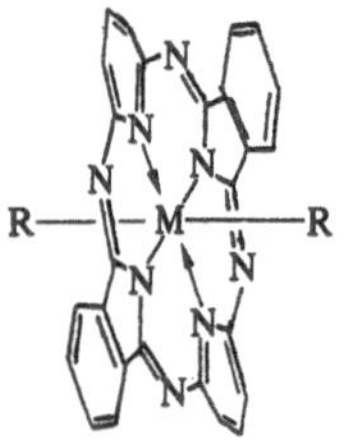

4.1 Narrow Arrangement of Bivalent N₄-Chelates in the Staple

4.1.1 Synthesis

Vacuum treatment of the dihydroxides (63)–(66) (R = OH) at 573–730 K led to the polyoxides (67) (Eq. 26)[104–113]. The water elimination is detected by thermal analysis[108–111]. Also heating the less soluble monomers in a heterogenous phase of high boiling solvents like 1-chloronaphthalene or quinoline is successful. With Si(OH)$_2$Pc (63) the corresponding polymer [Si(O)Pc]$_n$ (67) was prepared by heating in nitrobenzene with ZnCl$_2$[112]. Solvent and temperature must be considered, otherwise decomposition may occur.

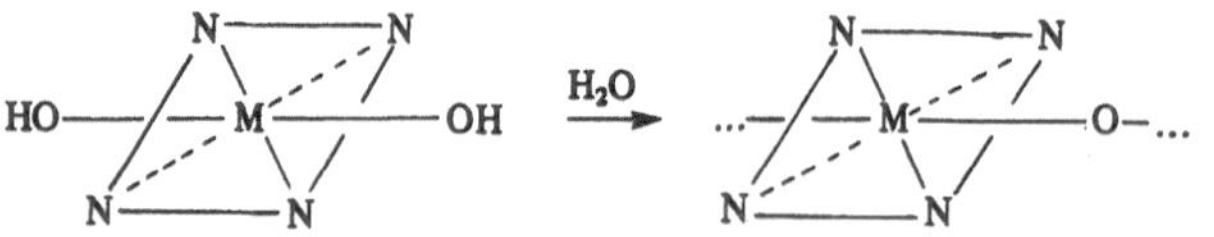

$$\text{HO} - M - \text{OH} \xrightarrow{\text{H}_2\text{O}} \cdots - M - \text{O} - \cdots \qquad (26)$$

(63)–(66)

(67) [Si, Ge, Sn(O)Pc]$_n$
     [Ge, Sn(O)Tpp]$_n$
     [Ge(O)EtioP]$_n$
     [Ge, Sn(O)Hp]$_n$

Table 8. Typical IR absorptions indicating dehydration of dihydroxides (63)–(66) to polyoxides (67)

Ligand	Metal	MO–H (cm^{-1}) of (63)–(66)	M--O-M (cm^{-1}) of (67)
phthalocyanine	Si	835–830	985–980
phthalocyanine	Ge	988	890–880
tetraphenylporphine	Ge	950	880–870
etioporphyrin	Ge	930	840
hemiporphyrazine	Ge	1050	890–865
tetraphenylporphine	Sn	835	850–840
hemiporphyrazin	Sn	955	840–820

In the IR spectra, the MO-H (stretch at 3630–3400 cm^{-1}, bend at 1050–830 cm^{-1}) of the monomers vanish and a new absorption typical for M-O-M (stretch at 985–825 cm^{-1}) occurs (Table 8). These absorptions are typical for octahedral surrounding (the Si-O-Si unit in tetrahedral configuration is absorbing at 1100–1000 cm^{-1})[108]. In the electronic spectra the polyoxides (67) show nearly the same absorptions as the inserted dihydroxides (63)–(66).

But sometimes the extinction value looses its intensity[112]. This was demonstrated by dehydrating Si(OH)$_2$Pc (63) in nitrobenzene (with ZnCl$_2$). During heating for 130 h at 448 K the intensity of the first intense and at 795 nm – VIS spectra in H$_2$SO$_4$ – decreased from ε_{max}^{mono} = 90.000 to ε_{max}^{poly} = 8420. It was assumed that the change in ε is directly correlated with the degree of polymerisation (Eq. 27) due to only one low energy electron jump in one polymer staple.

$$\bar{p} = \frac{(\varepsilon \cdot \delta)_{monomer}}{(\varepsilon \cdot \delta)_{polymer}} \tag{27}$$

It is postulated that the degree of polymerisation for [Si(O)Pc]$_n$ (67) obtained by vacuum treatment is 3 and by heating in nitrobenzene 11. But surprisingly no reduction of intensity was observed when going from Ge(OH)$_2$Pc to [Ge(O)Pc]$_n$[113]. This may be explained by vibronic coupling, which decreases with increasing distance between neighbouring Pc[114].

The connection of Pc through oxygen bridges was demonstrated by the reaction of Si(OH)$_2$Pc and Al(OH)Pc in a molar ratio of 1:2 in chloronaphthalene[115]. Small amounts of trimer (68) with n = 1 were detected after sublimation. Reaction in a molar ratio of 1:1 led to compounds with n > 1. The structure was determined only by elementary analysis, IR-spectra and hydrolysis.

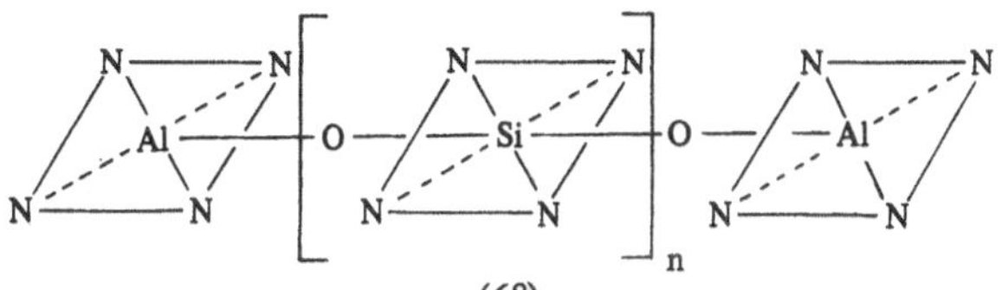

(68)

The kinetic of dissociation for $[Si(O)Pc]_n$ and $[Ge(O)Pc]_n$ was examined in conc. H_2SO_4[112, 113]. The kinetic stability of the polymers was enhanced with increasing degree of polymerisation. But the dissociation mechanisms are quite different: for $[Si(O)Pc]_n$ first order; for $[Ge(O)Pc]_n$ second order in respect to H_3O^+ concentration. So different intermediates were postulated (Eq. 28). This may be correlated with increasing stability of onium-ions of the IVa-group elements with higher atomic number.

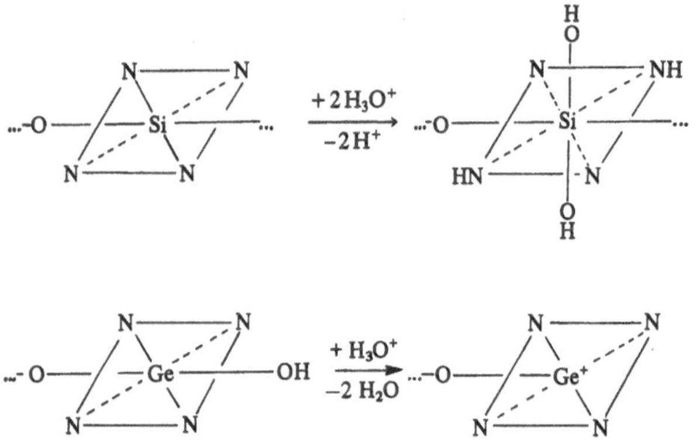

$$(28)$$

$$(29)$$

The black coloured μ-oxo-bridged Ni-tetraaza (14) annulene (*69*) was formed by heating $Ni(OH)_2Taa$ (oxidized NiTaa) at 473 K in bulk (Eq. 29)[11]. But the structure was only confirmed by elementary analysis and the amount of eliminated H_2O.

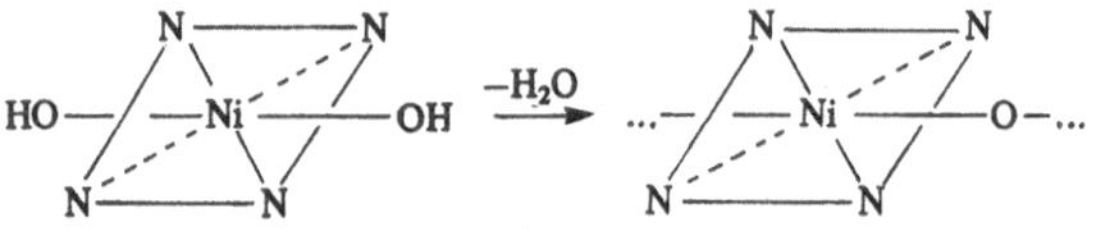

The Si, Ge, Sn containing N_4-chelates (*63*)–(*66*) react with monovalent alcohols, phenols and carboxylic acids to the corresponding alkoxides, aroxides and carboxides[108–111]. Polymers were prepared with analogous bivalent comonomers.

Heating $PcSi(OH)_2$, $Ge(OH)_2Pc$, $Ge(OH)_2HP$ with ethylenglycol in a molar ratio of 1 : 1 in high boiling solvents, polyoxyethylenoxy chelates (*70*) where synthesized. In order to get analogous polymeric porphyrin complexes, bifunctional $Ge(OCH_2CH_2OH)_2Tpp$, EtioP (*71*) where heated under vacuum[108–111] (Eq. 30).

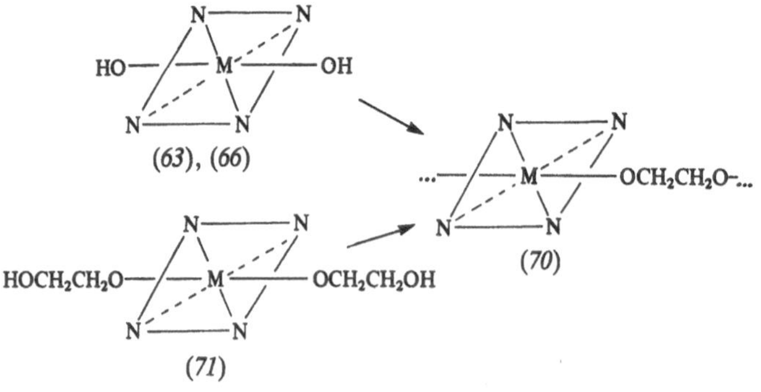

$$(30)$$

The reaction of $Si(OH)_2Pc$ with 1,2-bis(hydroxymethyl)carborane at 500–670 K or in H_2SO_4 was examined[116]. But only pyridine soluble oligomers were isolated.

All mentioned low molecular dihydroxides (63)–(66) are reacting with bivalent phenols like hydroquinone in solvents to the polyoxyphenylenoxy chelates (72) (Eq. 31)[108–111]. Also the dichlorides $Si(Cl)_2Pc$, $Ge(Cl)_2Pc$, $Ge(Cl)_2TPP$, $Ge(Cl)_2HP$ (63), (64), (66) can be used as monomers in some reactions.

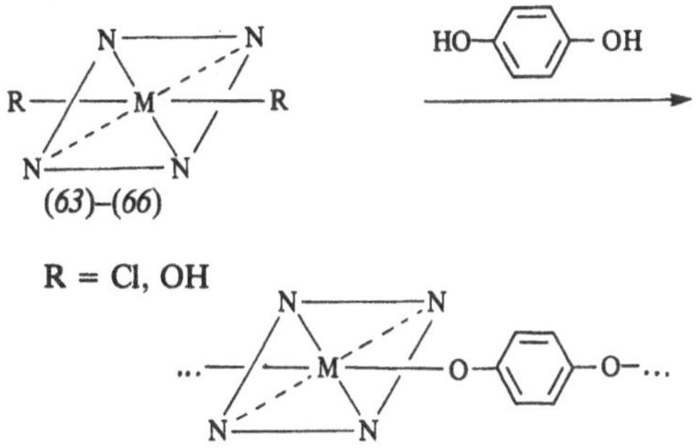

$$\text{(31)}$$

R = Cl, OH

Dicarboxylic acids as adipic acid were converted with $Ge(OH)_2Pc$, $Ge(OH)_2TPP$, $Ge(OH)_2HP$ (63), (64), (66) in bulk to give the polyoxyadipoyloxy chelates (73) (Eq. 32)[108, 109].

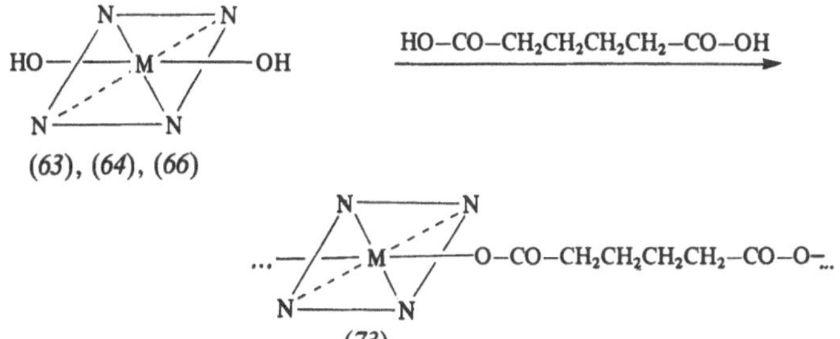

$$\text{(32)}$$

All mentioned polymeric oxides, alkoxides, aroxides and carboxylates are intensely coloured and insoluble in organic solvents. The structures were estimated by elementary analysis, IR spectra and model reactions. The mechanism of the reactions with alcohols and phenols is described as nucleophilic substitution at the metal atom for silonium and germonium ions[108, 109, 111].

A new kind of interesting polymer with an acetylene dianion between the metal atoms of N_4-chelates was synthesized[117, 118]. Polymeric chelates $[Si(C)_2Pc]_n$ (74) were prepared by reacting $Si(Cl)_2Pc$ (63) in THF with bis-bromomagnesiumacetylenes or $Si(C_2MgBr)_2Pc$ (75) (Eq. 33).

The dark green coloured polymers are unsoluble in organic solvents. The structures were confirmed by Raman spectra ($C{\equiv}C$ at 2024–2150 cm^{-1}), acid hydrolysis and model reactions with monovalent acetylenes[117–119].

Phthalocyaninato (μ-pyrazine)iron(II) (76) – a stable polymer with *coordinative link-age* – was obtained by reacting β-FePc with an excess of pyrazine in chlorobenzene

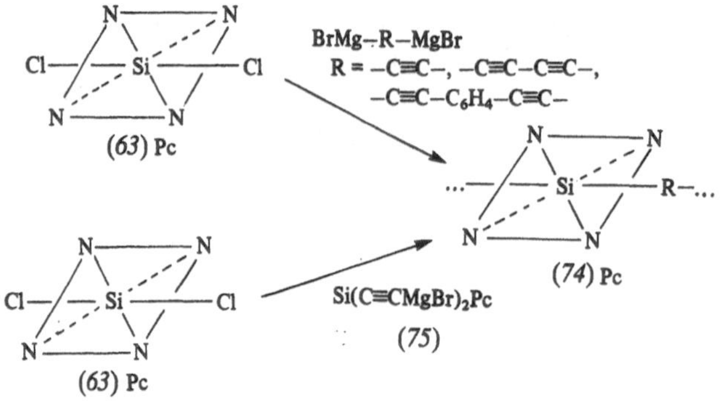

$$(33)$$

(Eq. 34)[118, 120, 121]. Surprisingly, new absorptions in solution at 788, 723, 714 nm with high O.D. values and in solid state at 770 nm indicate an intensive interaction[122]) between the Pc in the staple. Low molecular Fe(pyrazine)$_2$Pc, however, is absorbing at $\lambda \sim 660$ nm. Also with p,p'-bipyridyl and 1,4-diisonitrilobenzene polymers were prepared[121].

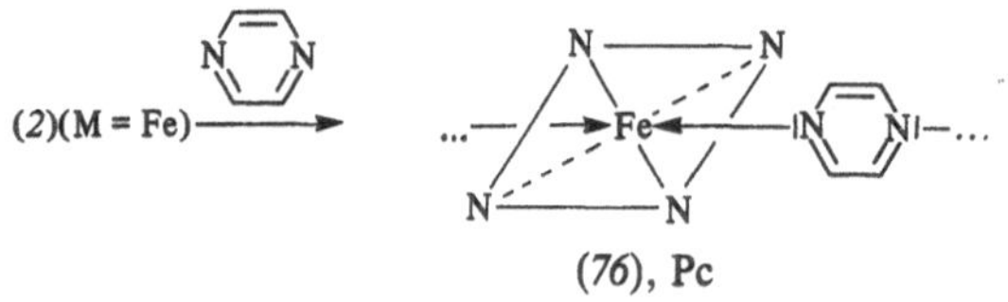

$$(34)$$

Polyoximes (77) with the molecular weight of 10^4–10^5 were obtained by interfacial synthesis of different dioximes (solved in NaOH) and metallocene dichlorides (solved in chloroform) (Eq. 35)[123]. The solubility of polymers depends on chain length and asymmetry of R'.

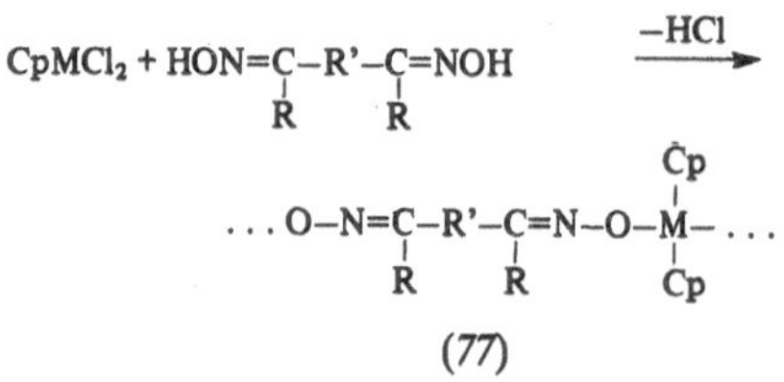

$$(35)$$

Dark green coloured imidazolate-bridged metalloporphyrin polymers (78) with combined covalent-coordinative linkage are synthetisized by reacting Mn(III)ClO$_4$-tetraphenylporphine with (Bu$_4$N)-imidazole in THF or methanol[124]. According to X-ray analysis, parallel polymeric chains contain long-long/short-short alternation of axial Mn-N$_{(Imid.)}$ bond lengths. Also with some dimers the antiferromagnetic coupling (measured by magn. susceptibilities) were calculated to be due to natural metalloproteins with magnetically interacting metal sites.

Evaporating Al(OH)Pc and Ga(OH)Pc with aqueous HF led after heating to fluoroaluminium and fluorogallium phthalocyanine polymers [Al(F)Pc]$_n$, [Ga(F)Pc]$_n$ (79)[125–128]. The polymers were purified by sublimation. The bond between the Pc in the staple is described as *combined covalent coordinative linkage*.

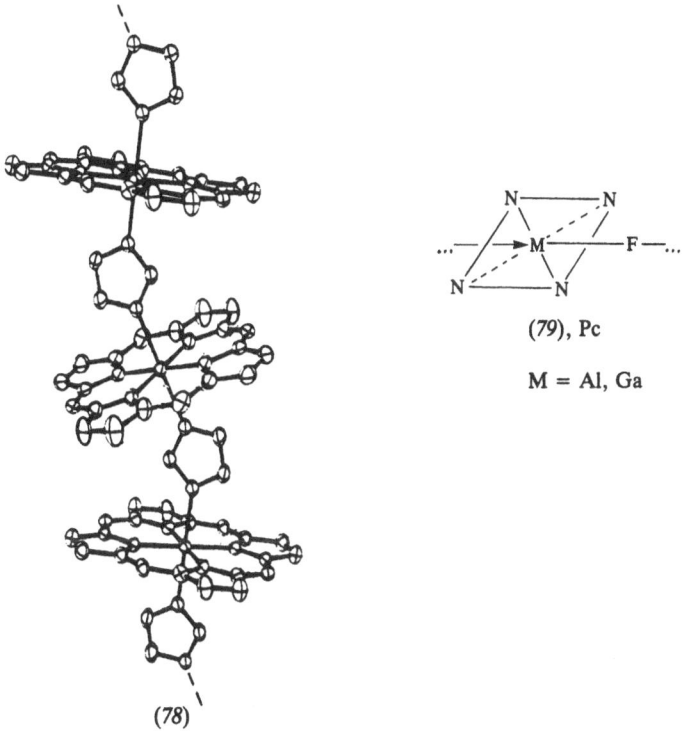

(79), Pc

M = Al, Ga

(78)

For a neodymium(III)-Pc with the formula [Pc₂Nd] H a new hydrogen bonded polymer structure in the crystal solid state is proposed[129].

4.1.2 Properties

Thermal stabilities of the polyoxides, -alkoxides, -aroxides and -carboxides (67), (70), (72), (73) were examined[109, 110]. In general, the stability is not enhanced drastically when comparing the polymers with low molecular model compounds. 10% weight loss in air occurs at 710–770 K. Investigations on *electrical conductivity* are more important.

Some results about polyoxy, polyalkoxy and polyphenoxy chelates are summarized in Table 9[108, 110, 130]. In principle, the Pc-chelates have better conductivity than the Hp-, Tpp-chelates (loss of aromaticity in Hp- and steric hindrance in Tpp-chelates compared with Pc-chelates). The polycondensation of dihydroxydes (63)–(66) to polyoxides (67) led to an increase in conductivity. Increase of distance between the Pc units in the staple results in decrease of conductivity due to diminishing intramolecular charge transfer[130]. Doping with electron acceptors like o-chloranile enhanced the conductivity due to charge transfer interaction to the donor Pc[130]. Only with [Sn(O)Pc]ₙ doping leads to an unaccountable decrease.

Doping of the polyoxides (67) with iodine leads to a surprising high increase in conductivity (Table 9)[131] by reduction of iodine; Raman spectra show the presence of $J_3^{\ominus}$ or $J_5^{\ominus}$. The measurements were done with compressed tablets, since matallic conductive single crystals of [(SiPcO)Jₓ]ₙ have not yet been obtained. Conductivity decreases from

Table 9. Dark conductivities (compressed powders) of some N_4-chelates containing IVa-group elements and oxygen in the chain

N_4-chelate[a]	No	$\sigma_{298}\,(\Omega^{-1}\text{cm}^{-1})$	$\Delta E\,(eV)$
$Si(OH)_2Pc$	(63)	$6 \cdot 10^{-9}$	0.79
$[SiPcO]_n$	(67)	$3 \cdot 10^{-7}$	0.80
$[(SiPcO)CA_{0.1}]_n$	(67)	$2 \cdot 10^{-3}$	0.17
$[(SiPcO)J_{0.5}]_n$	(67)	$2 \cdot 10^{-2}$	
$[(SiPcO)J_{1.4}]_n$	(67)	$2 \cdot 10^{-1}$	0.04
$[SiPcOC_6H_4O]_n$	(72)	$1 \cdot 10^{-12}$	1.45
$[SiPcOC_2H_4O]_n$	(70)	$4 \cdot 10^{-11}$	1.40
$[GePcO]_n$	(67)	$5 \cdot 10^{-10}$	1.40
$[(GePcO)CA_{0,1}]_n$	(67)	$1 \cdot 10^{-5}$	0.26
$[(GePcO)J_{2,0}]_n$	(67)	$1 \cdot 10^{-1}$	
$[SnPcO]_n$	(67)	$2 \cdot 10^{-6}$	0.6
$[(SnPcO)CA_{0,1}]_n$	(67)	$3 \cdot 10^{-9}$	1.0
$[(SnPcO)J_{5,5}]_n$	(67)	$2 \cdot 10^{-4}$	0.68
$[GeHpO]_n$	(67)	$2 \cdot 10^{-16}$	1.3
$[(GeHpO)CA_{0,1}]_n$	(67)	$9 \cdot 10^{-15}$	0.81
$[TppGeO]_n$	(67)	$3 \cdot 10^{-13}$	0.92
$[(TppGeO)CA_{0,1}]_n$	(67)	$4 \cdot 10^{-10}$	0.53

[a] Doping with o-chloranile (CA) or iodine (J). Index indicates molar ratio CA, J/N_4-chelate

Si < Ge < Sn because the distance of Pc rings in a staple is increased (Si–O–Si: 3.33 Å; Ge–O–Ge; 3.51 Å; Sn–O–Sn: 3.95 Å).

A new concept for metallic conductivity in staggered N_4-chelates was introduced by Hanack and Seelig[132, 133]. After calculating MO, a tetraazaporphyrine macrocycle with iron as central metal and acetylide ion (C_2^{2-}) as bridging ligand leads to a change in energy band structure.

The energy band should be partially occupied by overlapping of the originally filled valence band with one of the originally empty conduction bands of similar energy. Then

Table 10. Dark conductivities (compressed powders) of phthalocyanines connected by acetylide, -pyrazine (pyz) or fluoride

Pc	No	$\sigma_{298}\,(\Omega^{-1}\text{cm}^{-1})$	$\Delta E\,(eV)$
$Si(C{\equiv}CH)_2Pc$	(75)	$1 \cdot 10^{-11}$	
$[Si(C{\equiv}C)Pc)]_n$	(74)	$2 \cdot 10^{-11}$	
$Fe(pyz)_2Pc$		$2 \cdot 10^{-12}$	
$[Fe(pyz)Pc]_n$	(76)	$2 \cdot 10^{-5}$	
$[Al(F)Pc]_n$	(79)	$< 10^{-7}$	
$[(Al(F)Pc)J_{3.4}]_n^a$		0.59	0.03
$[(Al(F)PcJ_{2.4}]_n^{a,\,b}$		0.63	0.03
$[(Al(F)PcJ_{3.3}]_n^c$		3.4	0.017
$[Ga(F)Pc]_n$	(79)	$4.6 \cdot 10^{-10}$	0.85
$[(Ga(F)Pc)J_{2.1}]_n^a$		0.15	0,04

[a] unsublimed material doped with J_2, solid vapor reaction
[b] heating iodated sample two weeks in vacuo
[c] sublimed material doped with J_2 in CCl_4

metallic conduction must occur in the direction of the staple axis. But as mentioned in 4.1.1, silicon containing polymers (74) with analogous structure were obtained. Between the low molecular (75) and the polymer (74) the conductivity is not changed (Table 10)[118]. With pyrazine as bridging ligand the conductivity increases from the low molecular $Fe(pyz)_2Pc$ to the polymer $[Fe(pyz)Pc]_n$ (76) (Table 10)[118, 121]. Doping with iodine led to an increase of σ to 10^{-2} ohm^{-1} cm^{-1}. Also p,p'-bipyridyl and 1,4-diisonitrilobenzene bridged polymers show higher σ than the corresponding low molecular Pc[121]. More detailed results are expected after considering the original concept.

Doping unsublimed and sublimed stacked $[Al(F)Pc]_n$ and $[Ga(F)Pc]_n$ under various conditions with iodine leads to organic solids of high conductivity (Table 10)[127, 128]. The Raman spectra show that doping results in the reduction of iodine to J_3^{θ}. It is disadvantageous, that these polymers evolved iodine when stored at T = 300 K. Moreover, the iodine doped (67) is indefinitely stable in air; J_2 may be removed by heating above 373 K. The advantage of polymer (79) is the possibility to prepare thin films by sublimation.

4.2 Bifunctional N_4-Chelates as Comonomers with Other Polymers

$Si(OH)_2Pc$, $Ge(OH)_2Pc$ (63) and also $Al(OH)Pc$ were incorporated in polyesters by reaction with the central metal atom[134]. Two preparative ways were used taking polyethyleneadipate and terephthalate:
– polymer binding of the N_4-chelate during polyester formation
– polymer binding of the N_4-chelate at the polyester

The electronic spectra show that the phthalocyanines react with the carboxylic acid and not with alcoholic groups by forming carboxylates (Eq. 36).

$$\text{~~COOH} + \text{HO}-\underset{(63)}{\overset{\begin{array}{c}N\text{——}N\\M\\N\text{——}N\end{array}}{M}}-\text{OH} + \text{HOOC~~} \tag{36}$$

$$\xrightarrow{-H_2O}\quad \text{~~COO}-\underset{}{\overset{\begin{array}{c}N\text{——}N\\M\\N\text{——}N\end{array}}{M}}-\text{OOC~~}$$

By using a molar ratio of monomers (or polymer unit)/Pc = 10^5–10^4, it was found that the N_4-chelates are quantitatively bound into intensively coloured polyesters. Generally, addition of phthalocyanines during polycondensation leads to a slow decrease in molecular weight. However, by adding Si, $Ge(OH)_2Pc$ after polycondensation an increase of molecular weight was observed; this was due to chain connection of bifunctional N_4-chelates. Thermal properties and hydrolytic stabilities are not influenced through the covalent N_4-chelates. Also other chelates such as $Ge(OH)_2TPP$ (64) can be used. In principle, it was shown that covalent incorporation is also possible during polymerisation of ε-caprolactam and polyaddition reactions[134].

Polymers containing carborane and tetravalent SiPc as (80) were investigated for thermal stability[135].

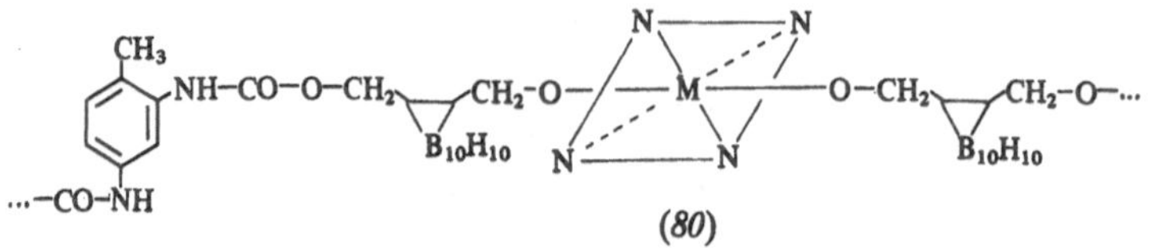

(80)

When heating (3°/min) in air up to 1170 K three regions of decomposition depending on the kind of sample were observed:
– Weight loss (5%) up to 620 K (dissociation of methane groups);
– Weight gain from 570–870 K (oxidation of oxymethylcarborane groups).
– Weight loss (18%) between 770–1170 K (thermal oxidation).

Polysiloxanes (82) containing tetravalent SiPc were prepared by reactions of SiPc-disilanole (81) with bis(ureido)siloxanes in xylene in high yields (Eq. 37)[136].

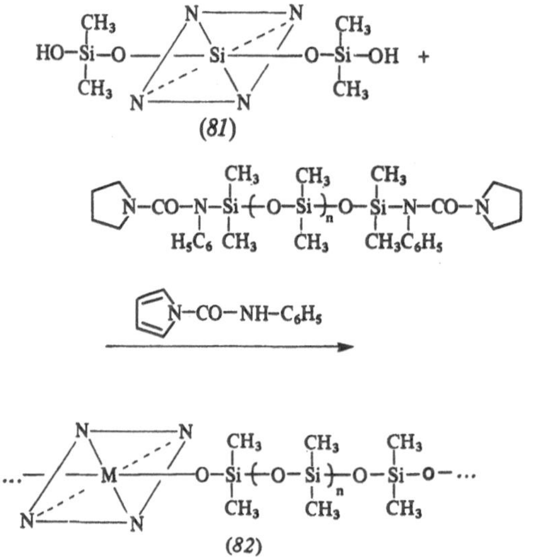

(37)

The polycondensation was studied in the ^{1}H-NMR. The dominant shielding effect of the phthalocyanine ring causes a clear separation of methyl group in the adjacent siloxane chain. The degree of polymerisation is 11–14 and was determined from the intensity ratio of terminal and internal Si-CH$_3$ groups. Since these polymers are readily soluble in organic solvents, films were prepared having melting points between 338–373 K.

5 Polymeric N$_4$-Chelates Through the Ligand

5.1 Polymeric Porphyrins

Only few informations are available on polymeric porphyrins in contrast to expected interesting properties such as visible light energy conversion, catalytic activity due to multiple porphyrin centers in vivo systems. The known preparative methods start from low molecular porphyrins.

Ni(II)-meso-tetramethylporphine (*83 a*) was converted by two ways into polymeric porphyrins[137, 138]. Firstly, bromation of (*83 a*) in CCl_4 in the presence of AIBN gave the expected monobrommethylporphyrin (*83 b*)[137]. This reactive intermediate (detected as methoxymethylderivate (*83 c*) is reacting easily with a β-pyrrol position of another porphyrin (Eq. 38). Beside dimer and trimer formation the polymer (*84*) was obtained (yield 21%). (*83 c*) is also converted into (*84*) with HCl (51% yield). In the electronic spectra the broader Soret band at 425 nm of (*84*) is shifted to the bathochrome side due to connection in β-position of the pyrrole ring compared with starting compound (*83 a*).

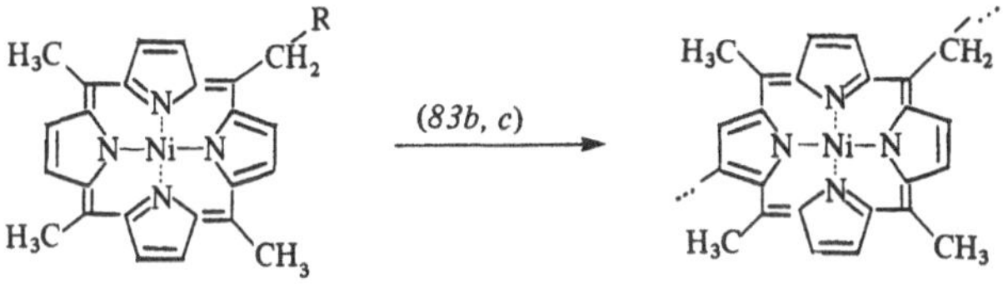

$$(83) \xrightarrow{(83b, c)} (84) \tag{38}$$

(*83*) R

a – H
b – Br
c – OCH_3

The second known possibility is the reaction of (*83 a*) with aldehyde or acetales under acidic conditions in $CHCl_3$ (Eq. 39)[138]. The mechanism of the condensation is analoguous to the reaction of phenols with aldehydes. Condensation at the β-position of (*83 a*) leads over dimer and trimer to the polymer (*85*) (yield < 30%). The yield of (*85*) may be optimized using a solvent with a higher solubility for the polymer. The limitation is the low yield of 4.3% for the one step synthesis of (*83 a*).

The violet coloured polymers (*84*) and (*85*) were investigated in the visible spectra, mass spectra and ^{1}H-NMR in relation to structural analogous dimers and trimers[137, 138].

$$(83a) \xrightarrow{R'-C\overset{H}{\underset{O}{\lessgtr}}} (85) \tag{39}$$

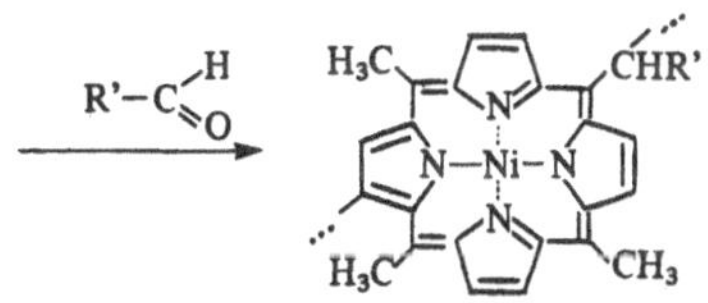

R = –H, –$(CH_2)_{10}$–CH_3

It is surprising that no polymers from substituted porphyrins such as (*9*) were described in detail. Only dimers and trimers were obtained from substituted porphyrins[139].

5.2 Polymeric Phthalocyanines (polyPc)

Two reviews summarize the literature until 1971[1, 140]. So, newer results are mainly discussed while elder works are relevant only under the present point of cognition.

The well known synthesis of low molecular phthalocyanines (*Pc*, *2*) starts from phthalic acid derivatives like 1,2-dicyanobenzene, 1,3-diiminoisoindolenine and phthalic anhydride[1, 5-8]. The yield of metal free or metal containing Pc is often high (80–100%). By starting with a bifunctional material like 1,2,4,5-tetracyanobenzene (TCB) or pyromellitic dianhydride (PMDA) a polymeric phthalocyanine (polyPc) (*86*) must be formed under the same conditions (Eq. 40). But the *determination of structure and molecular weight is very difficult.* Byproducts may be formed and the resulting polymers are often less soluble. But structure investigations are very important to correlate structure and property.

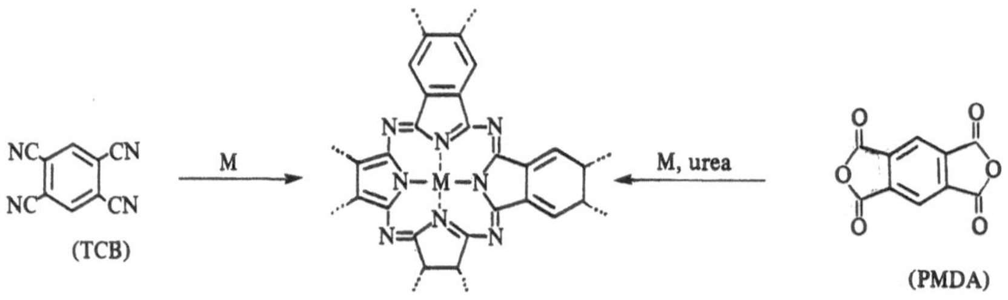

$$(86) \text{ (PolyPc)} \hspace{5cm} (40)$$

5.2.1 Simple Polymerisation of Nitriles to Polynitriles

Since the reactivity of the C≡N-group (Scheme 6) is important for polymer formation[1], some recent results are shortly mentioned:

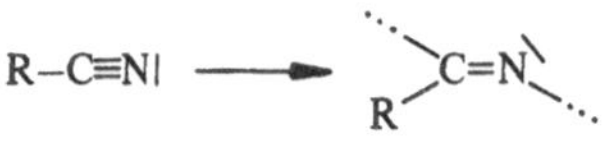

Scheme 6

– Poly(methinimin) (⁅HC=N⁆) formation through cationic ring opening polymerisation of 1,3,5-triazine[141].
– Electronic and molecular structure of ⁅HC=N⁆ polymers[142].
– Anionic, cationic and Ziegler-Natta type polymerisation of succinonitrile including polycondensations[143].
– Anionic polymerisation of fumaronitrile with butyllithium at 273–313 K in quantitative yield[144].
– Addition polymerisation of nitriles using organotin catalysts like tributyltin methoxide at 333–403 K[145, 146].
– Cationic and anionic polymerisation of 4-cyanopyridine at 493–593 K[147].
– Polymerisation of bis-4-cyanphtalimide with ethylzinc-acetanilide as initiator[148].
– Electroinitiated polymerisation of benzonitrile[149].

– Cationic polymerisation of benzonitrile in high electric fields[150].

– Copolymerisation of benzonitrile and propylene oxide with n-butyllithium[151].

Actually, in some cases the anionic polymerisation with reactive nitriles is possible under mild conditions. However, the cationic polymerisation needs higher reaction temperature. The same behaviour was found with TCB as reactive nitrile. The study of polymerisation of nitrile is of further interest in order to estimate side reactions with TCB.

5.2.2 Synthesis of Polymeric Phthalocyanines

5.2.2.1 Polymers from 1,2,4,5-Tetracyanobenzene (TCB)

TCB was converted in bulk with metal powders (Na[152], Mg[153], Cu[154]) or metal salts (Sb[152], Fe(III)[152, 155], Cu(I)[152, 154, 159], Cu(II)[154, 160], Zn(II)[152], Co(II)[161], Sc(III)[161], Zr(IV)[161]) and in high boiling solvents with metal salts (Cu(I)[154], Cu(II)[154, 155, 160, 162], Ni(II)[159, 160], Fe(II)[162], Fe(III)[155], Mo(II)[162]) to dark coloured metal containing polymers. In some cases also metal free polymers were isolated either starting from TCB or its methanol/NH$_3$ adducts[160, 163–165, 175] or after demetallisation under acidic conditions[152, 153]. Metal plates (Ti, Fe, Co, Ni, Cu) were covered with small amounts of gaseous TCB to homogeneous polyPc coatings[166, 174].

In order to study the polyPc-formation it is important to *isolate some intermediates*. Reacting TCB with Li-propylate as anionic initiator in propanol various reaction products were obtained (Eq. 41)[167]. With TCB and Li-propylat in a molar ratio ≤ 1 at 293 K the bisalkoxyisoindolenine (*87*) was isolated. It was converted at 338 K into the monocyclic tetraalkoxyisoindolo porphyrazin (*88*). By dropping a very dilute Li-propylat solution to a TCB solution (complete molar ratio TCB/Li-propylat ≥ 0.25) the reaction through the intermediate (*89*) led to octacyanphthalocyanine (*90*). Considering polyPc of a pure structure with only C≡N-end groups, the compound (*90*) is the first important intermediate for the polyPc-formation. After comparing all experimental facts (*87*) obtained at low temperatures is the product of the kinetic controlled reaction, while (*90*) obtained at high temperature is that of the thermodynamic controlled reaction. Like TCB the monocyclic (*90*) is converted with metal salts in solution or in bulk also to polyPc (*86*)[154, 173].

Another important feature is the *structural uniformity* of polyPc. Ortho di- or tetranitriles may be converted to poly-isoindolenines (polynitriles) as (*91*) and polytriazines as (*92*)[163, 168–172]. Generally, polymers prepared from TCB may contain the structure elements (*86*), (*91*), (*92*) (Eq. 42).

Moreover polyPc may exist in a two-dimensional parquet-floor (*86*) or a one-dimensional ribbon structure (*93*). No method is known to decide exactly between both structures. Mainly (*86*) and (*93*) in different ratios will be present in a polyPc. Only the combination of an absolute method of molecular weight and endgroup determination may solve this problem.

The problem of structural uniformity is discussed using copper as metal component. From TCB with Cu-powder and CuCl$_2$ or (*90*) with Cu(II)-salts (under strictly anhydrous conditions) uniform polyPc (*86*) was isolated[154, 173]. The IR-spectra of (*86*) are nearly in agreement with that of (*90*) (Fig. 6; A, B). Typical absorptions at 1500, 1305, 1090, 1025

(41)

(42)

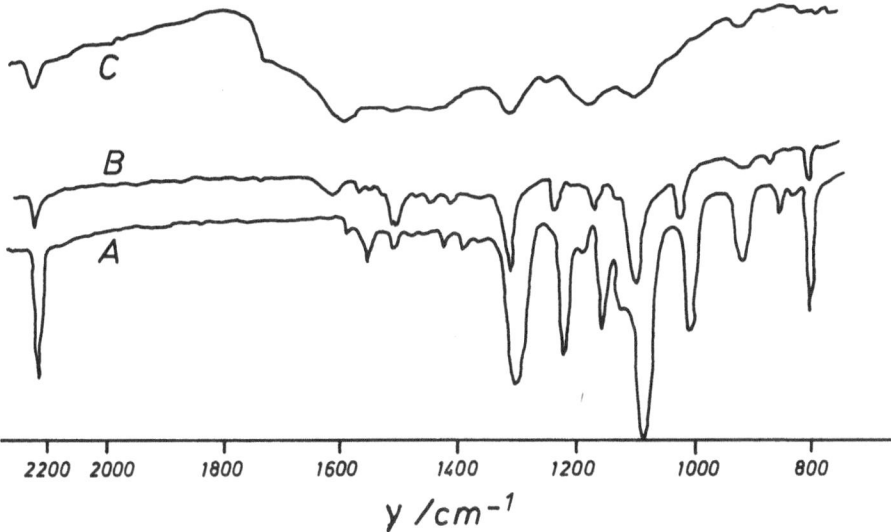

Fig. 6. IR-spectra (in KBr) of: A (*90*); B structural uniform polyCuPc (*86*); C structurally unpure polyCuPc containing (*91*)

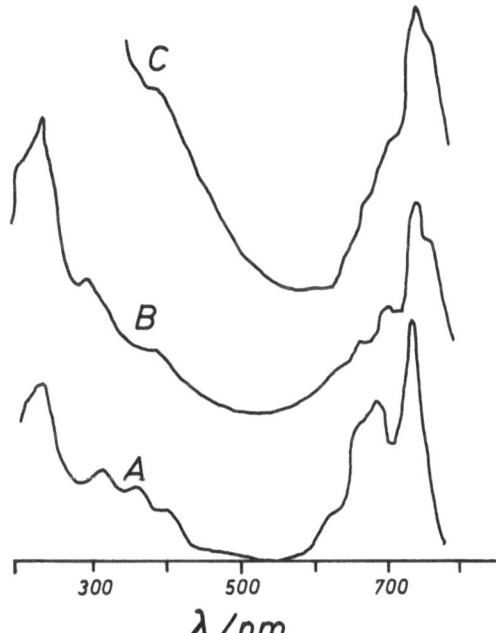

Fig. 7. UV/VIS-spectra (in conc. sulfuric acid) of: A (*90*); B structural uniform polyCuPc (*86*); C structurally unpure polyCuPc containing (*91*)

and 790 cm^{-1} indicate the complete metalation of the polymer (metal free poly Pc absorbs at 3260 (N–H), 1495, 1300, 1090, 1075 sh, 1020, 990, 675 cm^{-1}[175)].

A small absorption at 2225 cm^{-1} and no absorption near 1730 cm^{-1} proves nitrile and no carboxylic end groups. In the UV/VIS-spectra typical absorptions of (*90*) are present in (*86*) (Fig. 7, A, B). No long wave transposition is observed which may be expected for

π-electron delocalisation in the Pc-layers. On the other side a reaction between TCB and Cu(II)-acetylacetonate formed a polymer showing broad absorptions in the IR-spectra (Fig. 6, C). Also the UV/VIS-spectra have an very intensive absorption beginning at 500 nm and increasing with shorter wave length which is typical for polynitrile like (91) (Fig. 7, C). Pc-units are identified only at small absorptions at ~735 nm. There is no indication of triazine structure elements (92) having absorption lines at 1610, 1540, 1450 and 1370 cm^{-1}.

By comparing different reactands the following order of structural uniformity of (86) was established with TCB and copper[173]: CuCl$_2$, CuCl > Cu > Cu(II) acetate > CuSO$_4$ > Cu(II) acetylacetonate.

By taking TCB and Cu powder or CuCl$_2$ the electronic balance for (86) is equilibrated. For Cu(I) salt or other Cu(II) salts it is still unclear what will be oxidized when the dianionic phthalocyanine ring is formed.

By heating TCB with various metal powders the reactivity is decreasing in the following order (consumption of TCB at 673 K after 24 h 55–90%)[166]:

Cu ≫ Ni > Fe > Co

TCB with Cu$_2$Cl$_2$ was heated mainly with urea in bulk. Also copolymers from TCB and 1,2-dicyanobenzene were investigated[157]. The IR-spectra reveal that the fraction unsoluble in conc. H$_2$SO$_4$ has untypical broad absorptionbands. After changing reaction time, temperature and pressure IR and elementary analysis showed that best results with TCB and Cu$_2$Cl$_2$ are obtained at 573 K, 30 h, normal pressure (in closed ampoules), afterwards vacuum treatment.

If TCB was converted with Mg-powder and urea, the elementary analyses are in good agreement with the ribbon structure (93)[153]. But IR background absorption and C=O absorption indicate some unpurity in the polymer. Handling with 5% HCl or conc. H$_2$SO$_4$ led to saponification of free C≡N groups and demetallisation.

The reaction products of TCB with metal salts in ethylene glycol[155, 160] were investigated in detail[162].
The following procedure was described:

TCB + metal salt (Cu(II), Fe(II), Mo(II)) $\xrightarrow[\text{463 K, 5 h}]{\substack{\text{heating in}\\\text{ethylene glycol}}}$ isolation of unpure polymer

$\xrightarrow[\text{acetone, water 0,01 n HCl}]{\text{1. Purification: treating with}}$ removal of starting material $\xrightarrow[\text{with KOH/H}_2\text{O, 373K, 10 h}]{\text{2. Purification: saponification}}$

After isolation through acidification, IR and elementary analysis were used to identify the structure. The first purification left all polymers containing different end groups. In the case of Cu(II) the second purification lead to octacarboxy- or tetraimidoPc (94, 95). So under these reaction conditions the main reaction product is not a polymer. Comparing the metals both, incorporation during synthesis and complex stability increase in the order Mo < Fe < Cu.

By preparing metal free polymers it is shown that the reactivity to get high molecular compounds decreases using TCB > (90) > oligomers of (86) as starting materials[175].

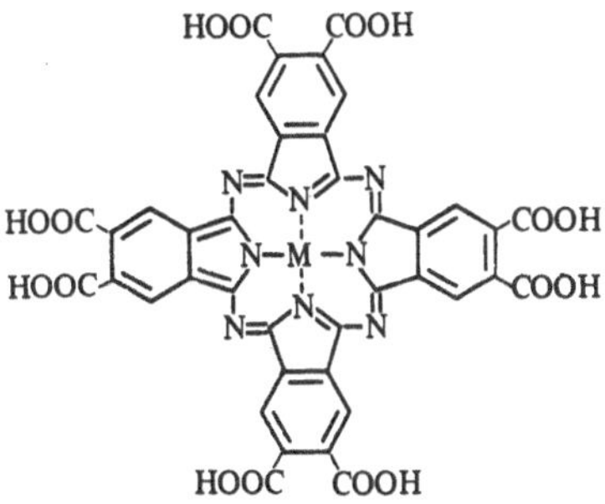

(94) 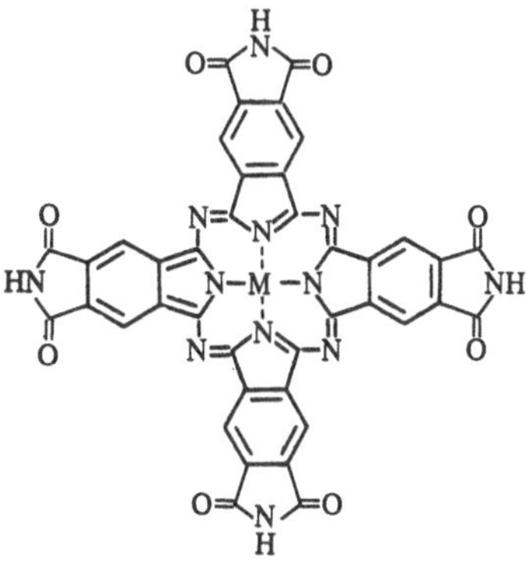   (95)

The conversion of TCB to metal free polymers needs some urea as initiator[164]. The obtained polymers were divided into acetone soluble, DMF soluble, DMF insoluble and conc. H_2SO_4 insoluble fractions. The highest yield of products soluble in organic solvents (50% acetone or 75–80% DMF soluble) were obtained with 1,5–3 mole % urea at 573 K in 10 h. DMF insoluble fractions resulted at 573 K in 40 h with an urea concentration > 20 mole %. In every case elementary analysis is in good agreement with expected values. But the IR-spectra show some C=O absorption (saponification) and broad background absorption (possible polynitrile (91) formation). An expected absorption typical for metal free Pc is present at ~ 1000 cm^{-1}.

Important is the comparison of methods for *molecules weight determination*[164] (Table 11). But some points are not considered:
– hydrolysis of the end groups will lead to some decomposition (15–25%[173]),
– uncertain formula for determination by IR.

Difficulties for obtaining exact molecular weights also occur when taking the metal free products of TCB and some metal disulfides[165]. After GPC diagrams the molar masses correspond to ~ 10^6 (Table 11). But from membrane osmometry, values near $2 \cdot 10^4$ were obtained. It was assumed that stiff Pc-structure will lead to a large effective hydrodynamic value which may explain the high GPC data.

5.2.2.2 Polymers from Various Tetranitriles

As reported earlier tetracyanoethylene, tetracyanothiophene, tetracyanofurane and other bis(1,2-dicyanoethylene-1,2-dithiolo)metal salts were converted to polymers[1, 140, 176]. New informations are not available yet. The structures of such polymers are not yet known.

New monomers like tetracyanotetrathiafulvalene (96) and tetranitriles (97) were examined[177–181]. (96) was converted with various metal acetylacetonates in bulk at 548 K to dark coloured polymers in yields of 61–87%[177]. The amount of incorporated metal is in good agreement with theoretical values. From different nitriles (97) polymers were produced by heating solely or in the presence of $SnCl_2$ or Cu (and partly using a nucleophilic initiator)[178–181]. Only in one case[181] the strucutre was determined. It was found

Table 11. Results from mol mass determinations (TCB: 1,2,4,5-tetracyanobenzene; PMDA: pyromellitic dianhydride, TCP: 3,3', 4,4'-tetracyanodiphenylether, DCB: 1,2-dicyanobenzene)

Method of preparation	Remarks	Degree of polymerisation	Method of mol mass	Ref.
TCP; Cu; 3 h; 548 K; bulk (partly with dicyanobenzene)	H₂SO₄ soluble	η = 0,03–0,12 g/100 ml	viscosimetry in H₂SO₄	176
TCB; TaS₂ or NbS₂; 24 h; 623–723 K; bulk	partly soluble DMF	$\bar{M} \sim 10^6$ and $\bar{M} \sim 10^4$ of some polymer	GPC (DMF) membrane osmometr.	165
TCB + DCB; 20 h; 573 K, bulk	H₂SO₄ soluble	$\eta \sim$ 0,04 g/100 ml	viscosimetry in H₂SO₄	157
TCB; CuCl₂ or FeCl₃; 5 h; 463 K; ethylene glycol	saponification KOH	octacarboxy CuPc (94)	elementary analysis	162
TCB; urea (200 mole%); 10 h, 573 K, bulk	acetone solub. fract.	$\bar{P} \sim 4$	ebullioscopy	164
	DMF solub. fract.	$\bar{P} \sim 6\text{–}7$ $\bar{P} > 6$	titration, COOH IR, C≡N	
	H₂SO₄ solub. unsolub.	$\bar{P} > 10$ $\bar{P} > 12$	IR, C≡N IR, C≡N	
PMDA; CuCl₂; 453 K; bulk	H₂SO₄ soluble	$\bar{M} \sim 1500\text{–}4000$	elementary analysis	185
PMDA; CuCl₂; 453–473 K; 0,5 h; bulk	saponification KOH	$\bar{P}$ = 1–5	elementary analysis	262
PMDA (partly with phthalic anhydride) different metal salts; 3 h; 448–483 K; bulk	H₂SO₄ soluble	$\bar{P}$ = 3–14	VIS in H₂SO₄; titration COOH	195
PMDA; FeSO₄; 5 h; 570 K; bulk	DMF soluble	$\bar{P} \sim 16\text{–}18$	GPC (DMF), Chromatogr.	193

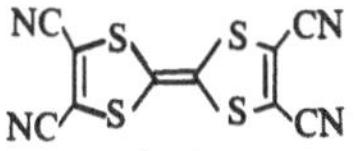

NC—S...S—CN / NC—S...S—CN

(96)

NC...R...CN / NC...CN

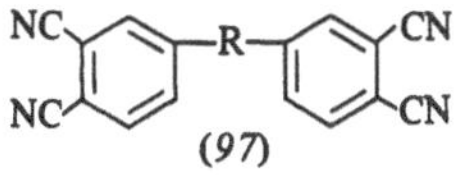

(97)

R = —N=CH–p–C_6H_4–CH=N–

—CF_2–O–CF_2–O–CF_2–

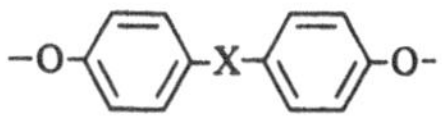

—O—⟨⟩—X—⟨⟩—O—

X = –, SO_2, $C(CH_3)_2$, $C(CF_3)_2$

—O—⟨⟩—O—

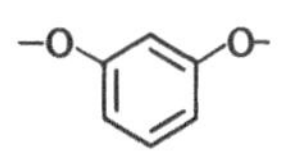

—NH–CO–$(CH_2)_x$–CO–NH–

x = 6, 10, 22, 36

that after heating a tetracyanocompound containing a bisphenol A bridge, the content of Pc in reaction mixtures is <2%. Addition of free phenol may lead to polymers with a higher extent of Pc-units[181]. So, some investigators must prepare Pc-polymers of pure structure.

5.2.2.3 Polymers from Pyromellitic Dianhydride (PMDA) and Derivatives

PMDA or the corresponding carboxylic acid partly with phtalic anhydride was converted with urea, catalyst (ammonium molybdate) and metal salts like $Cu(I)$[182, 183], $Cu(II)$[162, 184, 185], $Mg(II)$[153], $Zn(II)$[184], $Co(II)$[186], Fe[187], $Fe(II)$[188], $Fe(III)$[189, 190], $Ni(II)$[184], $Ga(III)$[184], $Ca(II)$[191] (Eq. 40). Metal free polymers were produced by treating the poly-CaPc with acidic solution[191]. On the other side it was shown that metal ions like $Cu(II)$, $Zn(II)$, $Fe(II)$, $Ni(II)$ may be introduced into metal free polymers very slowly[192]. PMDA is a cheaper starting material than TCB. Therefore the structure investigation is of high interest. It was tried to study the structure of the polymers in more detail using the following procedure[162]:

PMDA, $CuCl_2$, urea,   heating at 453–523 K   isolation of
ammonium molybdate   ——————————→   unpure polymer
        in bulk

    1. Purification with
——————————————→   isolation through acidification
   acetone, water, 6 n HCl
2. Saponification with KOH/H_2O

After elementary analysis and IR, all end products contain only carboxylic acid end groups; it is assumed that depending on the molar ratio of reactands, monomers (94) to pentamers were formed. After carrying out the elementary analysis, the oligomers are not quantitativly metallized by the described reaction.

PolyFePc prepared from pyromellitic acid, $FeSO_4$, urea and catalyst were investigated by Mössbauer spectra[193, 194]. These results allow to estimate the purity. The origi-

Table 12. Mössbauer spectra of polymeric iron phthalocyanine (δ experimental isomer shift, Δ quadrupole splittings, A relative areas of absorption lines, Γ line widths; details s. 193)

Phthalocyanine	δ (mm·s^{-1})	Δ (mm·s^{-1})	A	Γ (mm·s^{-1})	Interpretation
original polyFePc	0,01	2,95	0,31	0,56	low mol. polyFePc
	0,28	2,66	0,22	0,48	Pc units interior of the polymer
	0,07	1,97	0,34	0,48	Pc units peripheral of the polymer
	0,30	0,83	0,13	0,58	oxidized polyFePc
chromatographed polyFePc	0,34	2,53	0,24	0,51	Pc units interior of the polymer
	0,08	1,99	0,63	0,51	Pc units peripheral at the polymer
	0,36	0,85	0,13	0,48	oxidized polyFePc
polyFePc after heating 570 K (decarboxylation)	0,32	2,73	1,00	0,42	Pc units interior and peripheral at the polymer
low mol. FePc	0,31	2,59	1,00	0,22	

nal polymer (after reprecipitation from H_2SO_4) shows four superimposed quadrupole doublets (Table 12)[193]. After chromatography over Sephadex LH 20 (to exclude oligomeric Pc and impurities with mol masses lower than 5000) and heating to 570 K (desorption of O_2, decarboxylation of COOH-end groups) only one quadrupole doublet $\Delta = 2.7$ mm/s is present[193]. A similar quadrupole splitting was observed earlier[194].

VIS-spectra and titration of carboxylic end groups helped to get mol masses of polymers prepared from pyromellitic acid and various metal salts (Table 11)[195]. The degree of metallation is quite variable and ranges between 90% (with $CuCl_2$) and 50% (with $ZnCl_2$)[184]. The polymers exhibit similar VIS-spectra as the low molecular Pc[184]. Therefore strong π-electron delocalisation from one tetraazaporphyrin ring to the neighbouring ones may not be favoured. But it is typical to find marked reduction in λ_{max} accompanied by broadening of the absorption bands. Quantitative statements are not possible because the polymers contain impurities[162].

5.2.2.4 Polymers from Various Tetracarboxylic Acid Derivatives

As reported earlie 3,3', 4,4'-tetracarboxydiphenylether or -sulfon and others were used as starting materials analogous to PMDA[182, 196, 197]. In one case osmometry showed a mol mass of ~ 12.200 ($\overline{P} \sim 13$)[196b].

In order to get more structurally pure polyPc a new way may start from multifunctional low molecular Pc. Co(II), Ni(II) and Cu(II) containing bis(dicarboxybenzoyl) Pc diimides were converted with various aromatic diamines in N-methylpyrrolidone to soluble polyamides which react to insoluble polyimides (98) by heating to 473 K[198]. Melt and solution condensation of (12 b) and 3,3'-diaminobenzidine led to unsoluble polymers (99) with Pc-rings linked by benzimidazole units[199].

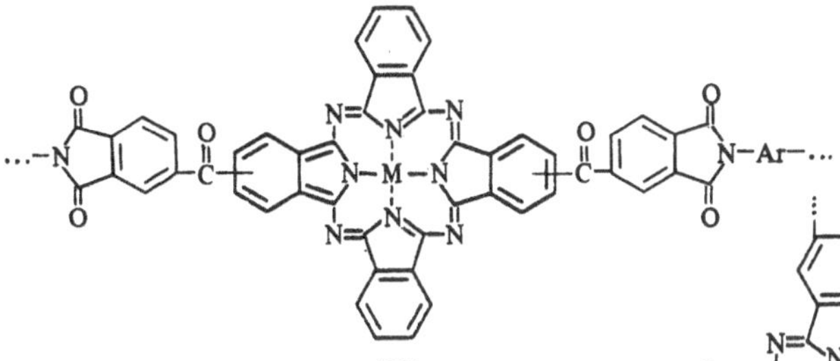

(98)

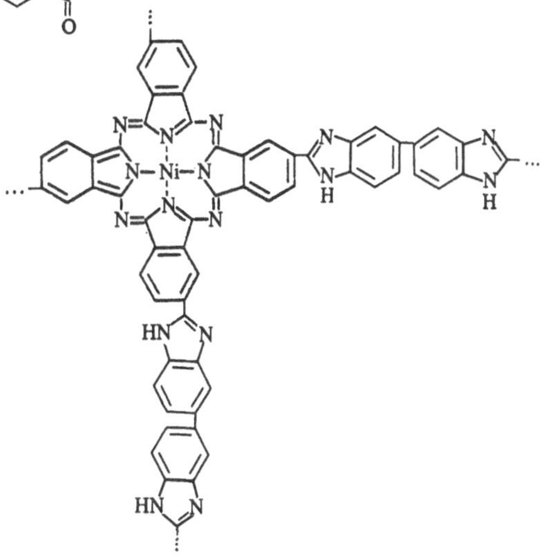

(99)

5.2.2.5 General Properties

Polymers obtained from different methods are black, blue or brown coloured compounds. The black or brown is probably due to the impurities. In principle of uniform polyPc structure exhibit nearly the same colour as the low molecular analogs. Impurities are easily determined by UV/VIS- and IR-spectra[154].

Independent of the impurity the solid state structure is influenced by the preparation: high crystallinity (<50–60%) by reaction in bulk under normal pressure or vacuum combined with additional heating[157, 158]; amorphous products by bulk reaction under high pressure or reaction in solution[158, 200, 201]. Polymers with uniform end groups are prepared from TCB, (90)[154, 166, 167] or PMDA[162].

The solubility depends to a great extent on the kind of Pc: low molecular unsubstituted CuPc is less soluble in org. solvents, well soluble in conc. H_2SO_4; octacyano CuPc (90) has good solubility; polymers differ in solubility fractions in org. solvents and conc. H_2SO_4 depend on the preparation procedure.

Polymers containing e.g. Co, Cu, Ni, Al are stable against treatment with conc. H_2SO_4 for a short time while those with Mg, Cd, Pb, Sn, Fe decompose to give metal free polymers. During this procedure end groups may be saponified. In general *hydrolytic stability* of the Pc-ligand ring under acidic conditions is lower than those of low molecular Pc. Metal free and metal containing polymers prepared from PMDA and partial with phthalic anhydride were investigated at various temperatures in 17 n H_2SO_4 for their stability[192, 202].

Investigations of *thermal stability* also lead to lower stability of polymers compared with low molecular Pc. The decomposition of polymers prepared from tetracyanobenzene[153, 17)], tetracyanothiophene[200)], tetracyanofurane[200)], tetracyanoethylene[202, 203)] starts in air at 523–573 K. Under inert atmosphere or vacuum the stability is 100–150 K higher. The stability is improved by heating the polymer after preparation under vacuum in order to eliminate impurities and some reactive end groups.

PolyPc (98), (99) prepared from low molecular substituted Pc show a better thermal stability[198, 199)], because side reactions are avoided during synthesis.

Moreover, polymers prepared from (97) with a stable or flexible bridge between ligand units exhibit a combination between thermal stability, adhesiveness, and mouldability[178–181)].

Such monomers with separated bifunctional sides for polymer formation show better flexibility of the polymer by the linkage between Pc-units. The heating of (97) produces a prepolymer gel; it can be processed into any desired form by additional heating. In air such polymers are stable over several thousand hours at 523–573 K.

5.2.3 Electrical Conductivity

The *dark conductivities* of the polymers were mainly measured on compressed powders (as tablets). Also few measurement are reported with thin films on metal plates[140, 166)]. Metallic conductivity was never observed as in the case iodine doped NiPc[204)] and [Ge(O)Pc]$_n$ (s. 4.1.2). In every case semiconducting behaviour according to Eq. 43 is typical for the polymers

$$\sigma = \sigma_0 \exp(\Delta E/2\,kT) \qquad \begin{aligned} \sigma &= \text{spez. conductivity} \\ \Delta E &= \text{thermal activation energy} \end{aligned} \qquad (43)$$

σ_{298} is in the order of 10^{-1}–10^{-13} ohm^{-1} cm^{-1} with $\Delta E \sim 0.1 - 1.5$ eV. σ depends on the surrounding atmosphere (doping, introduction of traps); it increases with increasing pressure of compressed powders (better packing, intermolecular charge transfer). Highly important for reproducible electrical properties are:

– development of standard procedures for preparation and purification (to get the same kind of polymer)
– degassing at higher temperatures with high vacuum (degassing of impurities, small molecules like O_2, H_2O; formation of more uniform end groups)[156, 159, 183)].

Since the structural uniformity of the polymers is uncertain the results on conductivity are only roughly comparable. The main results are summarized as follows:

5.2.3.1 Polymers from Tetracyanobenzene

The maximum σ_{298} of polyPc prepared with Cu_2Cl_2 in bulk is $10^{-1} - 10^{-2}$ ohm^{-1}cm^{-1}[152, 153, 155–159, 183, 205)]. After treatment with conc. H_2SO_4, soluble and unsoluble fractions show $\sigma_{298} \sim 10^{-3} - 10^{-4}$ ohm^{-1} cm^{-1}[157)]. Therefore acid containing groups decrease σ. The amorphous polymers (obtained with CuCl at 240 atm pressure of N_2) show lower σ_{298} of 10^{-5} than those with higher cristallinity of 50–60% (obtained with Cu_2Cl_2 under normal pressure and temperature treatment) with σ_{298} of 10^{-2} ohm^{-1}cm^{-1}[158)].

In thin films of polyCuPc prepared in-situ from TCB on Cu-plates in the gas phase, the conductivity decrease is not linear with film thickness[166]. This may be due to formation of less high molecular Pc because the Cu concentration is low on the reacting side.

In addition, the following results were described: thermoelectromotoric force (n-type conductors)[156, 159, 205]; influence of different metal atoms on σ of polyMPc[152, 153, 155, 159]; σ of metal free polymers[152, 163, 206].

Only few reports try to get some informations about mobility μ and number n of charge carriers[156, 207]. From the low mobility μ of $10-10^{-1}$cm^2 V^{-1}s^{-1} it may be assumed that the relative large value of σ corresponds to the fact that the generation of charge carriers requires no energy and that the motion may be activated with increasing temperature.

5.2.3.2 Polymers from Pyromellitic Dianhydride (PDMA)

In general the polymers prepared from pyromellitic acid derivatives have lower σ with higher ΔE-values.

Cu containing polymers exhibit σ_{298K} of $10^{-3}-10^{-8}$ ohm^{-1}cm^{-1} [155, 183, 207-209]. Gas elimination during heating and the influence on σ was studied in detail[183]. Lower σ was observed for polymers containing Fe ($\sigma_{298} < 10^{-8}$) and Co ($\sigma_{298} < 10^{-4}$)[189, 209].

5.2.3.3 Polymers from Other Bifunctional Monomers

Main data[1, 140] for polymers ($\sigma_{298} < 10^{-1}$) from other monomers are available on:
tetracyanoethylene:
   influence of metal atom[201, 210-214]; metal free polymer[202, 210, 215]; investigation of films on metal plates[213, 216]; p- or n-type conductivity[205, 206, 209] frequency dependence[213, 214, 217]; anisotropic conductivity[216];
tetracyanothiophen and others:
   comparison of tetranitriles; influence of metal atom, p-type conductivity[200, 218]

Polymers from tetracyanotetrathiafulvalen (96) and metal salts exhibit σ_{298} of 10^{-4} to 10^{-10} ohm^{-1}cm^{-1} [177]. Doping with tetracyanoquinodimethane has no influence on σ. This is surprising because doping of tetrathiafulvalene leads to the well known metallic conductivity.

Polymers from (97) are less conducting ($\sigma_{298} \sim 10^{-12}$ohm^{-1}cm^{-1})[178 a]. The heating of such polymers for 65 h at 870 K increases the value to 10^{-2}ohm^{-1}cm^{-1} due to pyrolysis (weight loss 18%).

Polyimides (98) have σ_{298} of about 10^{-6} to 10^{-11}ohm^{-1}cm^{-1} [198]. The temperature dependence of σ has a typical break point at 310-330 K with lower ΔE values at higher temperatures.

The *mechanism of the intermolecular*[132] *and intramolecular*[152] *conductivity* is still unclear. In general, the polymers show a higher conductivity than the low molecular unsubstituted Pc (2) (Table 13). Surprisingly the low molecular octacyano-Pc's (90) – the model compounds for structural uniform polyPc (86) – also show high σ-values $< 10^{-1}$ohm^{-1}cm^{-1} [219]. In (90) intramolecular charge transfer is not dominant; the conductivity must be described fairly well over a intermolecular charger transfer. In contrast to polyCu Pc and poly CoPc ΔE reaches for polyFePc (from PMDA) a limit value of

Table 13. Conductivity datas on some phthalocyanines (TCB: 1,2,4,5-tetracyanobenzene, PMDA: pyromellitic dianhydride). Measurements on compressed powders under vacuum

Compound	Preparation	σ_{298} (ohm^{-1}cm^{-1})	ΔE (eV)	Ref.
CuPc, NiPc, FePc	(2) –	$10^{-13} - 10^{-15}$	1.4–1.9	
OctacyanCuPc	(90) TCB	$2 \cdot 10^{-2}$	0.2	167, 173, 219
OctacyanNiPc	(90) TCB	$5 \cdot 10^{-2}$	0.2	167, 173, 219
OctacyanFePc	(90) TCB	$5 \cdot 10^{-8}$	0.7	167, 173, 219
polyCuPc, polyNiPc	(86) TCB	$< 10^{-2}$	0.15	s. text
polyCuPc	(86) PMDA	$< 10^{-3}$	0.3	s. text
polyFePc	(86) TCB	$< 10^{-3}$	0.3	152, 155
polyFePc	(86) PMDA	$< 10^{-8}$	0.8	209

> 0.8 eV[189, 209]. It was assumed that Fe is disturbing the intramolecular π-electron delocalisation (rubi-conjugation)[259]. Also the low molecular octacyanoFePc (90) has low σ and high ΔE-values[219]. However, according to other reports polyFePc exhibits $\sigma \sim 10^{-2}$–10^{-6} with $\Delta E \sim 0.15$–0.65[152, 155].

5.2.4 Catalytic Properties

The following situation may be generalized to a great extent: The oxidation of acetaldehyde ethylene acetal to ethylene glycol monoacetate (Eq. 44) is catalyzed by so-called "PolyMPc" (prepared from TCB and metal chlorides in ethylene glycol)[220]. Table 14 lists the activity using low molecular MPc and the so-called mono- and bimetallic PolyMPc[220]. The mentioned reaction products from metal chlorides, TCB in ethylene glycol seem to be low molecular tetraimidoPc (95) and never polymers[221]. Also activities of these Pc's prepared 1973 are quite different from those obtained 1968 (Table 14). Therefore reactions and activities described furtheron are qualitative true but the *quantitative side must* consider purity, uniformity, stability and turn over numbers of the catalysts.

$$CH_3-\underset{\underset{O-CH_2}{|}}{\overset{\overset{O-CH_2}{\diagup}}{CH}} \quad \xrightarrow{O_2} \quad CH_3-\overset{\overset{O}{\|}}{C}-O-CH_2-CH_2-OH \tag{44}$$

Instead of acetales also free aldehydes may be oxidized[220, 222, 223]. The oxidation of acrolein is carried out at 313 K in benzene[222]. For polyCu- and polyFePc, pyridine is needed to activate the oxidation. From benzaldehyde and others a mixture of benzoic acid and peroxybenzoic acid is produced with Cu- and FePc derivatives (95) (prepared from TCB and metal chlorides in ethylene glycol) (Eq. 45)[220 c]. The reactions were carried out in various solvents at 303 K.

$$C_6H_5-CHO + O_2 \xrightarrow[\text{cat.}]{} C_6H_5-COOOH \tag{45}$$

$$\xrightarrow[\text{cat.}]{C_6H_5-CHO} 2\ C_6H_5-COOH$$

Table 14. Oxidation of acetaldehyde ethylene acetal by Pc as catalyst (50 mg catalyst, 5 ml substrate, 303 K)

Pc-derivative		Amount of O_2 (ml) absorbed after 6 h	Ref.
Cu(II)Pc	(2)	little	220
Fe(II)Pc	(2)	little	220
Fe(II)Pc	(2)	235	221
subst. CuPc from TCB[a]	(95)	0	220
subst. FePc from TCB[a]	(95)	0	220
subst. FePc from TCB[a]	(95)	~ 100–257	221
subst. Cu/FePc from TCB[a]	(95)	~ 0–320	220
subst. Cu/FePc from TCB[a]	(95)	~ 200	221

[a] prepared in ethylene glycol

The mechanisms for the oxidation of different aldehydes in the presence of Fe- or Cu/FePc-derivatives in solvents at 283 K were investigated[223]. Starting from acetaldehyde the peracetic acid is produced quantitatively within 2 h. This is surprising because low molecular Pc may decompose peracetic acid. The rate equation (Eq. 46) shows a strong dependence of O_2-consumption on catalyst concentration.

$$-\frac{d(O_2)}{dt} = k\,(CH_3CHO)^{3/2}\,(catalyst)^{1/2}\,(O_2)^{1/2} \tag{46}$$

According to the general opinion the initiation process is hydrogen abstraction by the activated oxygen bound at the catalyst (Eq. 47)

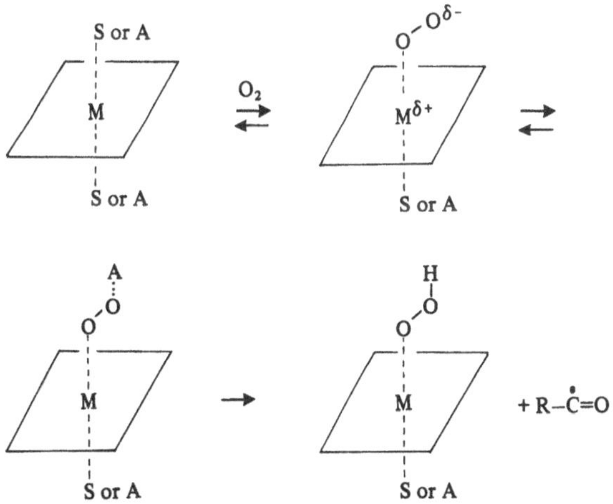

$$(47)$$

The stability of the peroxy-group is also demonstrated by heating cumenehydroperoxide at 353 K in p-xylene with polyPc prepared from TCB and metal chlorides in bulk[224]. Surprisingly purified polyPc does not decompose the hydroperoxide, whereas only crude polymer does it completely. This again demonstrates the need to use pure polymers in order to get reproducible results. Cumene is oxidized at 353 K to the mentioned hydroperoxide with pure polyPc[224]. Bimetallic polymers, such as polyCu/FePc, are more active then the polyCuPc or polyFePc. Interestingly the induction period using polyCu/FePc decreased from 100 min to 15 min with small amounts of pyridine activating the interaction of the catalyst with O_2 (Fig. 8).

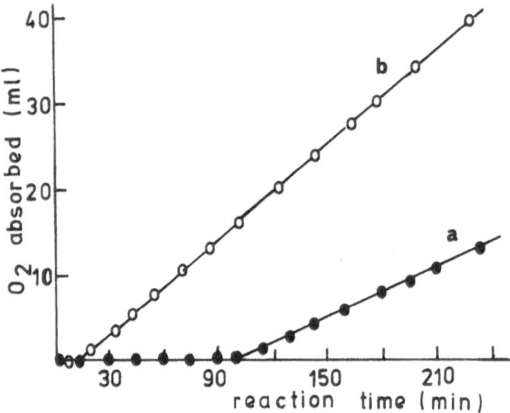

Fig. 8 a, b. Oxidation of cumene (0.652 M) to cumene hydroperoxide using polyCu/FePc (4.82 $\cdot$ 10^{-3} M in xylene at 355 K). (a) without pyridine, (b) addition of pyridine (0.225 mol/l)

In principle, the ability of the polymers to catalyse mild oxidations of alkylaromatic compounds by forming hydroperoxides at a secondary or tertiary carbon atom was also shown before[225].

The oxidative polymerisation of 2,6-dimethylphenol to poly[oxy(2,6-dimethyl-1,4-phenylene)] (Eq. 49) with various catalysts (100)–(102) was investigated in detail[226–228].

$$\text{(48)}$$

Polynuclear copper complexes containing maleonitriledithiol as ligand (100) (solid state structure see[229]) show high reactivtiy (Table 15). If some cupric-pyridine complex is added to the inactive complex with the cation $2 \cdot (CH_3)_4 N^+$, the activity is increased due to the exchange of $(CH_3)_4 N^+$ with the Cu ion. Polymers containing phthalocyanine or hemiporphyrazine units together with bis(ethylene-1,2-dithiolato) Cu(II) structure elements (101, 102) are less active. Addition of copper-pyridine complex leads to solubilization of the unsolved catalyst due to ligand exchange reactions (Eq. 49). Then catalytic activity is increasing significantly (Table 15).

It is said that the high activity corresponds to an arrangement in which the CuPc structure element is acting as electron-acceptor for the formation of an O_2-adduct while the dithiolo unit works as coordinating centre for the substrate. So a multi-electron transfer may be realized.

It is assumed that also the oxidative dehydrogenation of alcohols like methanol, ethanol and isopropanol with the polyPc goes through hydrogen abstraction at the oxygen bond. The polymers prepared from TCB and Cu_2Cl_2 in bulk were treated at ~ 393 K

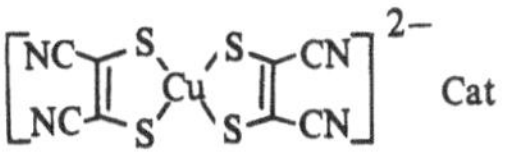

(100) Cat $= Cu^{2+}, 2(CH_3)_4N^+$

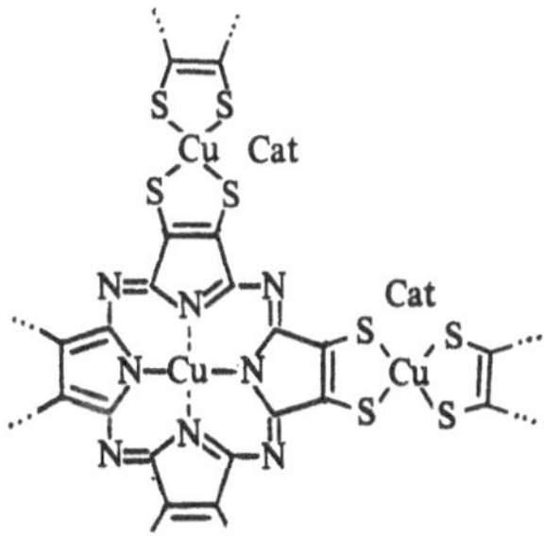

(101)

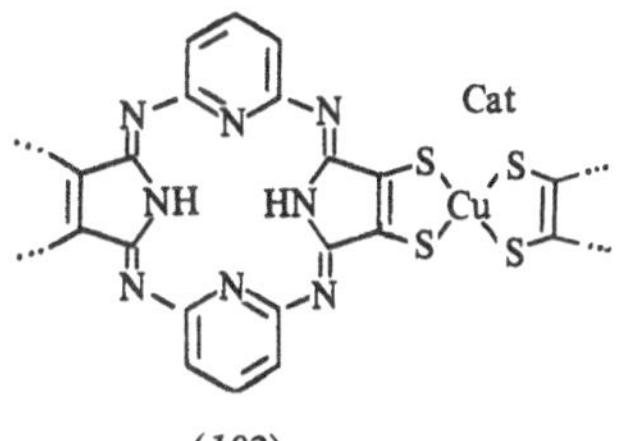

(102)

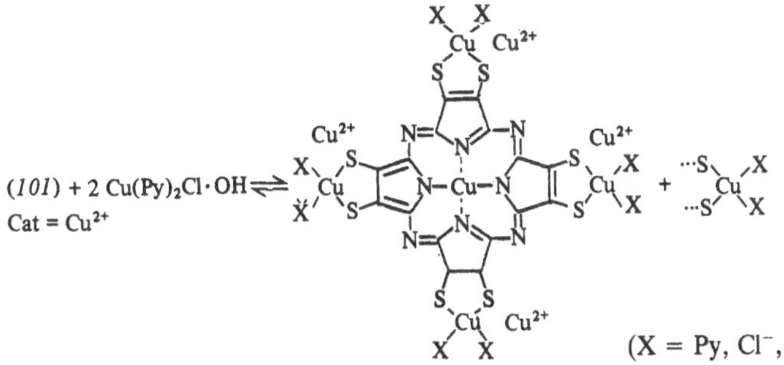

$(101) + 2\,Cu(Py)_2Cl\cdot OH \rightleftharpoons$ $(X = Py, Cl^-, OH^-)$ (49)

Cat $= Cu^{2+}$

Table 15. Catalytic activities of some Cu maleonitriledithiol complexes for the oxidative polymerisation of 2,6-dimethylphenol (0,5 M) at 303 K in pyridine

Complex		Concentration (g atom metal/l or polymer unit/l)	Addition of Cu(II)Py (mol/l)	O_2-uptake (μl/min/ml solution)
No	cation			
(100)	$2\cdot N(CH_3)_4^+$	0.005	–	0
(100)	$2\cdot N(CH_3)_4^+$	0.005	0.005	43
(100)	Cu^{2+}	0.005	–	61
(101)	$2\cdot N(CH_3)_4^+$	0.005	–	3.7
(101)	Cu^{2+}	0.5	–	3.5
(101)	$2\cdot N(CH_3)_4^+$	0.005	0.005	86
(101)	Cu^{2+}	0.1	0.005	33
(102)	$2\cdot N(CH_3)_4^+$	0.05	–	0
(102)	$2\cdot N(CH_3)_4^+$	0.05	0.005	62

with gaseous alcohols[230]. Beside the expected aldehyde or ketone also CO_2 and H_2O are formed. Surprisingly the activity decreases in the order of polymers containing:

Cu > Fe > Fe/Cu > Fe/Mn

This is different from the oxidation of aldehydes in the liquid phase. No explanation is known[230]. Low molecular FePc is much less active.

FePc-derivatives (*95*) prepared from TCB and FeCl₃ in ethylene glycol as mentioned before were used as electron transfer carriers for the reduction of α-substituted ketones and others with 1-benzyl-1,4-dihydronicotinamide or benzenethiol (Eq. 50)[231, 232].

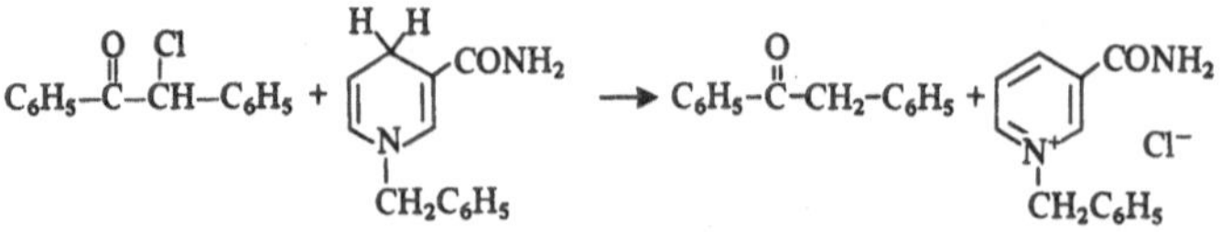

$$
\underset{\text{O Cl}}{C_6H_5-\overset{\text{O}}{\underset{\|}{C}}-\overset{\text{Cl}}{\underset{\|}{CH}}-C_6H_5} + \text{(pyridine-CONH}_2\text{, CH}_2C_6H_5\text{)} \longrightarrow C_6H_5-\overset{\text{O}}{\underset{\|}{C}}-CH_2-C_6H_5 + \text{(pyridinium-CONH}_2\text{, Cl}^-\text{, CH}_2C_6H_5\text{)}
$$
(50)

The reactions were carried out in benzene or aqueous methanol at 353 K. Mostly a high amount of catalyst was used (molar ratio ketone: Fe(II)ion = 1 to 14). The activity of the FePc derivative is higher than that of low molecular unsubstituted FePc or other MPc. The activity of the Fe containing catalysts may be due to redox reactions abstracting halogen atoms. Therefore the FePc's support the electron transfer from the donor to the acceptor (Eq. 51).

$$
\underset{\text{O X}}{-\overset{\text{O}}{\underset{\|}{C}}-\overset{\text{X}}{\underset{|}{CH}}-} + \text{Fe(II)Pc} \longrightarrow \text{Fe(III)Pc X}^\ominus + -\overset{\text{O}}{\underset{\|}{C}}-\overset{\bullet}{CH}-
$$

$$
\text{BNAH} + \text{Fe(III)PcX}^\ominus \longrightarrow \text{Fe(II)Pc} + \text{BNAH}^{\overset{+}{\bullet}}\ X^\ominus
$$
(51)

$$
\text{BNAH}^{\overset{+}{\bullet}}\ X^\ominus + -\overset{\text{O}}{\underset{\|}{C}}-\overset{\bullet}{CH}- \longrightarrow -\overset{\text{O}}{\underset{\|}{C}}-CH_2- + \text{BNA}
$$

Until now only reactions were mentioned using Pc as catalyst under mild conditions. At higher temperatures working with a continous flow system at atmospheric pressure the following reactions were studied with polyPc prepared from TCB and metal chlorides in bulk:

a) Oxidation of hydrogen sulfide to sulfur (Eq. 52) with order of catalyst reactivity[233]:

Fe > Co > Cu > Mn

$$
H_2S + 1/2\ O_2 \xrightarrow[\substack{433-473\ K}]{} \underset{<70\%}{S + H_2O}
$$
(52)

b) Dehydrogenation of ethylbenzene to styrene (Eq. 53) with order of catalyst reactivity (Fig. 9)[234]:

Cr > Ni > Co > Fe, Cu

$$
C_6H_5-CH_2-CH_3 \xrightarrow[\substack{773-823\ K}]{} \underset{<13\%}{C_6H_5-CH=CH_2 + H_2}
$$
(53)

c) Hydrodesulfurization of aliphatic thiols to alkanes (Eq. 54) with order of catalyst reactivity[235]:

Pt, Pd, Cr > Ni > Co > Fe, Cu

$$
C_2H_5SH + H_2 \xrightarrow[\substack{623-723\ K}]{} \underset{<100\%}{C_2H_6 + H_2S}
$$
(54)

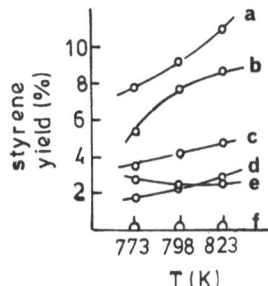

Fig. 9. Dehydrogenation of ethylbenzene to styrene over various polyPc (a. polyCrPc; b. polyNiPc; c. polyCoPc; d. polyFePc; e. polyCuPc; f. polyH$_2$Pc). $p_{styrene}$ = 0.1 atm; p_{N_2} = 0.9 atm

d) Decarbonylation of furfural to furane (Eq. 55) with order of catalyst reactivity [236]:

Pt > Pd > Co > Ni > Fe

$$C_4H_3OCHO + H_2 \rightarrow C_4H_4O + CO + H_2 \tag{55}$$

In the absence of oxygen polyFePc are less active (reactions b–d) while in the presence of oxygen as with reaction a) polyFePc is active. This was also often observed in other oxidation reactions mentioned before. In general in reactions a–d bimetallic polyPc show higher activity than monometallic ones. Including earlier mentioned results except some structural unsecurity, the combination of two metal centers seem to produce a combined catalytic effect.

The decomposition of H$_2$O$_2$[209, 237–239], N$_2$H$_4$[140, 240], HCOOH[241] and the activation of H$_2$ (p-H$_2$ → o-H$_2$; H$_2$ + D$_2$ → 2HD)[242] were studied in detail using various types of polyPc.

5.2.5 Electrochemical Properties

For electrocatalysis in *fuel cells*[243] the cathodic reduction of O$_2$ and the anodic oxidation of e. g. H$_2$, N$_2$H$_4$, CH$_3$OH needs active catalysts which must fullfill the following requirements: high open circuit cell voltage (lowering H$_2$O$_2$-formation); reduced polarisation in the working function of the cell; long time stability. The most common catalyst Pt is expensive and has the disadvantage to be active for the cathodic and anodic side. So mixed potentials may reduce the half cell voltage[243].

Square planar metal chelates such as MP (*1*), MPc (*2*), and Schiffbase chelates (*6*) reduce O$_2$ selectively as MTAA (*4*) for the anodic side[243, 244]. In Fig. 10 low molecular and polymer phthalocyanines (*2*), (*86*) (both precipitated on active carbon) are compared with Pt for the reduction of O$_2$ (Eq. 56). According to this, polyFePc (*86*) has nearly the same voltage and current density as Pt.

$$O_2 + 4e^- + 4H^+ \rightarrow 2H_2O \tag{56}$$

The long time stability of polyMPc ranges over several hundred hours[209, 245].

Active electrodes containing (*86*) are produced as follows (details up to 1974 see also[244]):

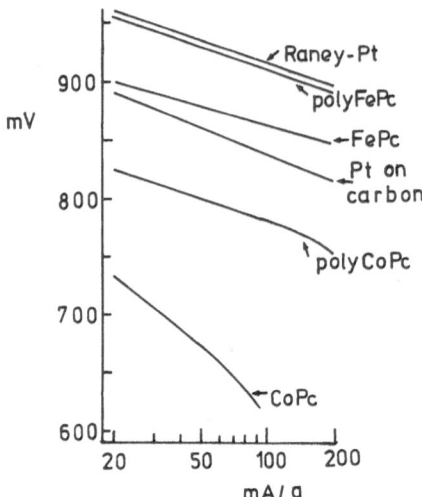

Fig. 10. Comparison of the catalysts/platinum, (2), and (86) for the O_2-reduction. Electrolyte: 4.5 n H_2SO_4

- Preparation from PMDA, metal salt and urea; solving in conc. H_2SO_4 in the presence of basic carbon and then precipitation of the supported catalyst with water; the coated powders containing 10–50% (86) made into porous electrodes bound with e.g. polyethylene[209, 245–250]. Polymers from PMDA were also pressed on gold gauze[246], sprayed from aqueous alcohol suspension e.g. on Ni-plates[249] or precipitated from pyridine or H_2SO_4 directly on a rotating disk electrode[251]. Polymers from PMDA on active carbon were also used as a dispersion electrode[251].
- In situ gas phase preparation directly on the carrier (carbon or graphit) either from TCB and a volatile chelate such as metal (Fe, Mn, Fe/Cu, Fe/Sn) dipivaloylmethanes at 520–670 K[249, 250, 252–255] or from PMDA and $FeCl_2$ in a flow of NH_3 gas[248]. Loading is 10–50%.

The unsecurity what kind of MPc is used for electrocatalysis led to the comparison of low molecular dimer and polymer FePc[255]. With elemental analysis only, it is impossible to estimate the polymer as hexamer. Also, the dimer structure, if only determined by N/ Fe elemental ratios, is very unsecure. Problems with various oligomers (partly metal free[251]), (90) and different end groups are not solved. One C 1s peak (ESCA-spectra) for the dimer[255] is unreliable due to different kind of benzene rings.

In general, the electrochemical measurements were carried out in a half cell arrangement (using a counter electrode and reference electrode) either under acidic (4,5 n H_2SO_4)[209, 245–248, 251, 253], neutral[253] or alkaline (1–6 n KOH)[249–251, 253, 254] conditions.

From the different notes the following may be estimated:

- The half wave potentials of low molecular MPc (2) and polyMPc (86) are in the same order[257]. Therefore the active centers in (86) are also active for O_2-reduction. The higher activity of polyMPc compared with MPc (Fig. 10) may be due to higher electrical conductivity (electron transport) of the polymer.
- Influence of pH-value[253].
- Increase of Fe content in polymers enhances activity[248].
- Increase of activity also by thermal treating (crosslinking may lead to higher conductivity)[248].

Table 16. Correlation between conductivity (σ), katalase activity and activity for O_2-reduction at 298 K

PolyFePc No.	σ_{298} (ohm^{-1}cm^{-1})	Catalase activity t_{50} (min)	Potentials for O_2 reduction at a current density of 20 mA/cm^2
1	$8 \cdot 10^{-12}$	25	530
2	$2 \cdot 10^{-11}$	23	645
3	$4 \cdot 10^{-10}$	5	700
4	$1 \cdot 10^{-9}$	3	720
5	$2 \cdot 10^{-8}$	0,6	795

– Most interesting are results correlating electrical conductivity, H_2O_2-decomposition and electrochemical activity for O_2-redcution[209]. In general better σ-values led to higher activity[245] both for the katalase reaction and the O_2-reduction as shown for polyFePc prepared from pyromellitic dianhydride (Table 16)[209]. The correlation between catalase activity and O_2-reduction may be due to a polarisation by H_2O_2 originated from O_2-reduction.

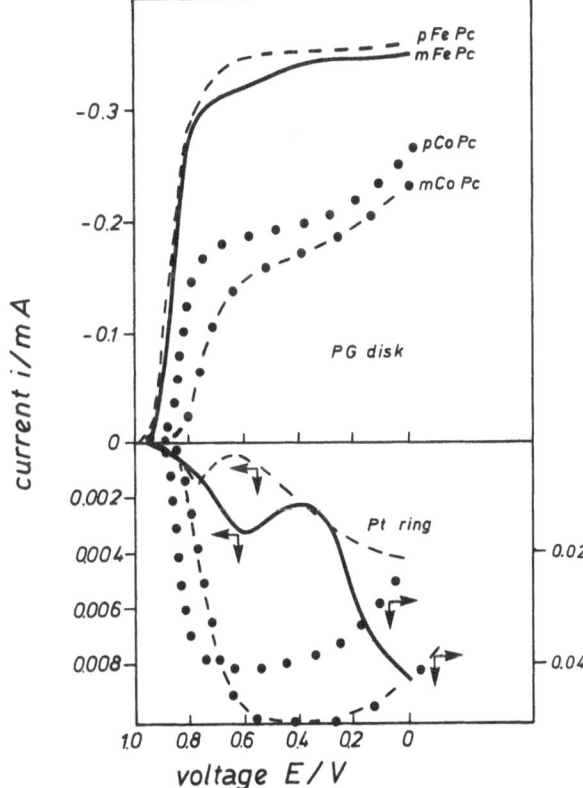

Fig. 11. O_2-reduction at low molecular and polymeric Pc (2), (86) supported at rotating ring disk electrode. Electrolyte: 1 m KOH. Upper curve typical for O_2-reduction to H_2O; below typical for O_2-reduction to H_2O_2

- Increase of activity in the order Cu < Co < Fe due to better electron donor function of the metal[209, 244, 245].
- Influence of preparative technique. Gasphase in situ synthesis rendered catalysts 2–3 times more active than those prepared by the dissolution-precipitation procedure[249].
- Effect of surface structure of support used[254].
- Highly important for the use of MPc and polyMPc is the fact that the four electron reduction of O_2 to H_2O is much more favoured than the two electron reduction to H_2O_2. Mechanistic studies at a rotating disk electrode shows low concentration of H_2O_2 in solution (Fig. 11)[251].

Coatings of polyMPc prepared in situ form TCB on metal plates were investigated by cyclic voltametric measurements[166, 174]. Such electrodes show stable anodic and cathodic current densities comparable with a Pt electrode. Additionally, a large photo current was observed when using coatings of polyMPc on Ti-plates[258].

5.3 Polymeric Hemiporphyrazines (Polyhexazocylanes)

One common possibility for the preparation of low molecular hemiporphyrazines is starting from o-dicyano compounds such as 1,2-dicyanbenzene and diamines such as 1,4-diaminobenzene or 2,6-diaminopyridine to obtain the ligand followed by introduction of metal ions (in the case using diaminopyridine)[1]. In order to get polymers, concepts of bifunctionality with various tetracyano compounds (*103*) and diamines (*104*) were used to prepare the polymer ligand (*105*); this is in some cases able to form polychelates (*106*) (Eq. 57, Table 17).

(*103*) (*104*)

(57)

(*105*) (*106*)

(*103 a*) (TCB) (*103 b*) (*103 c*) R′ = -SO₂- (*103 d*) R′ = -O- (*103 e*)

(*103 f*) (*103 g*)

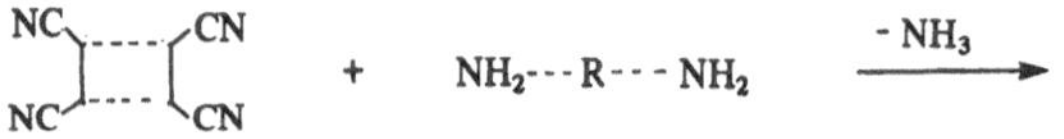

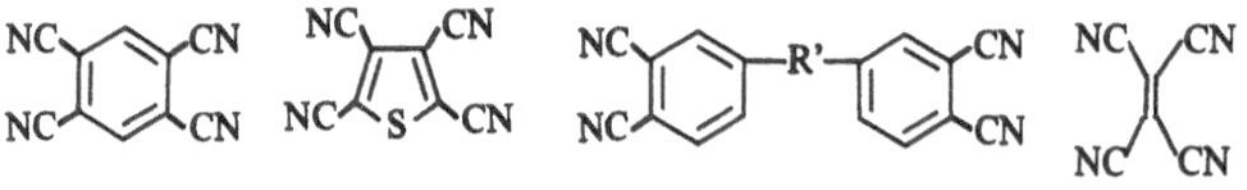

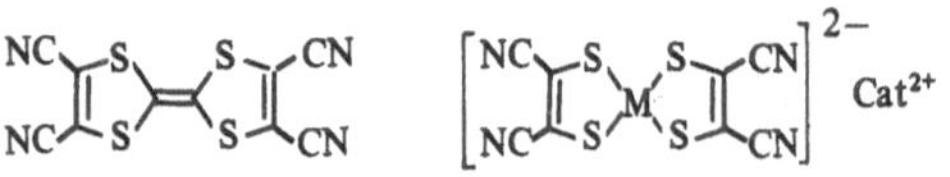

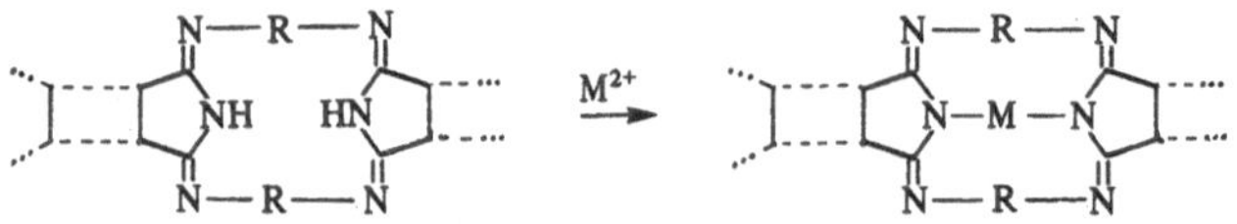

Table 17. Polymers from various tetranitriles (103) and diamines (104)

Tetranitril[c]	Diamine[c]	Conditions	Yield polymer/% (amount evolved NH₃/%)	Structure elements in polymer	Methods structure investigation	Solubility polymer[a]	Ref.
(103 a)	(104 a)	phenol/LiCl, 450 K	90(97)	(105) ≫ (110)	¹H-NMR,	+−	260
(103 a)	(104 a)	alcohols/alkoxides	73–96 (10–52)			+−	260
(103 a, c, d)	(104 a, f–i, k, l)	phenol/LiCl, 450 K, 860 mbar	60–95	(105) ≫ (110)	model reaction	+(−)	259, 261
b	(104 h)	N-methyl-pyrrolidone	92	(105)		+	259, 261
(103 a, b, e, g)	(104 a, e)	bulk, 473–573 K; 1-choronaphthalene	67–95	(105),	IR	−	264
(103 f)	(104 a, e, f, g, i)	bulk, 473 K; phenol, anthracene	55–75	(108)–(110)	IR	−	265
(103 a)	(104 a, b–e, g, h)	bulk, 600 K	30–95 (55–76)		IR	−	263
(103 a)	(104 h)	methoxyethanol/alkoxide	85	(105), (108), (109)	IR		263
(103 a)	(104 a, b, e, h, m)	methoxyethanol/alkoxide	75–92 (40–110)		model reactions		266
(103 a)[d]	(104 h)[d]	alcohols/alkoxides	~90 (small)	(104)	model reactions		268

a solubility e.g. DMF, DMSO, HMPT: + good, +− indifferent, − unsoluble
b 1,3-dimethoxy-5,7-diiminopyromellitdiimid
c molar ratio 1:2
d molar ratio 1:1

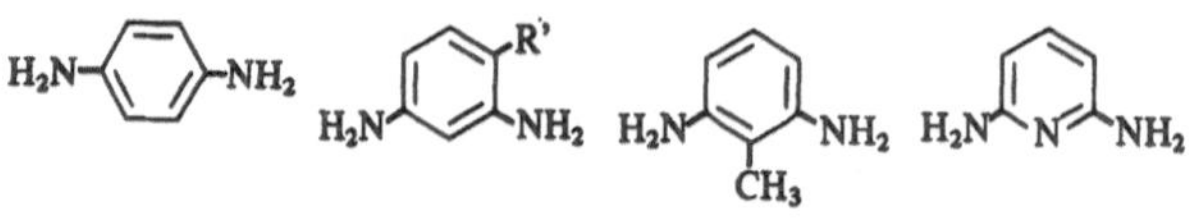

(*104a*) (*104b*) R′ = -H (*104d*) (*104e*)
(*104c*) R′ = $-CH_3$

(*104f*) R′ = -
g R′ = $-CH_2-$
h R′ = -O-
i R′ = -C(=O)-
j R′ = $-CH=CH_2-$
k R′ = $-SO_2-$

(*104l*) (*104m*)

Detailed investigation on the reaction mechanism and the structure of the products shows that the conditions of getting polymeric hemiporphyrazine depend extremely on the reaction conditions and kind of diamine. Starting from TCB (*103a*) polymers may contain the structure elements (*105*), (*107–110*) (Eq. 58). Normally the *limits for getting structurally pure* (*105*) are *very narrow* (Table 17).

(*103a*) TCB (*105*) (*107*) (*108*) (*109*) (*110*)

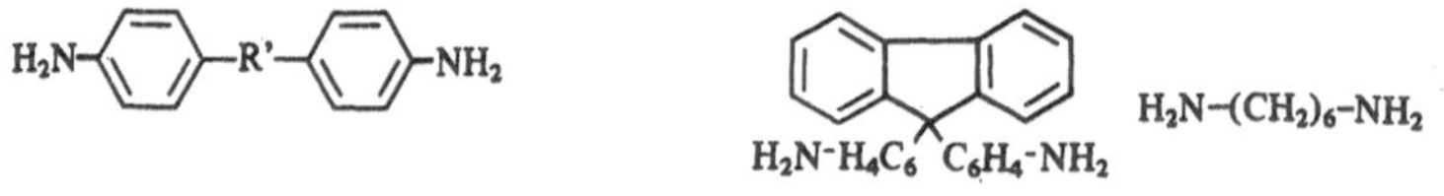

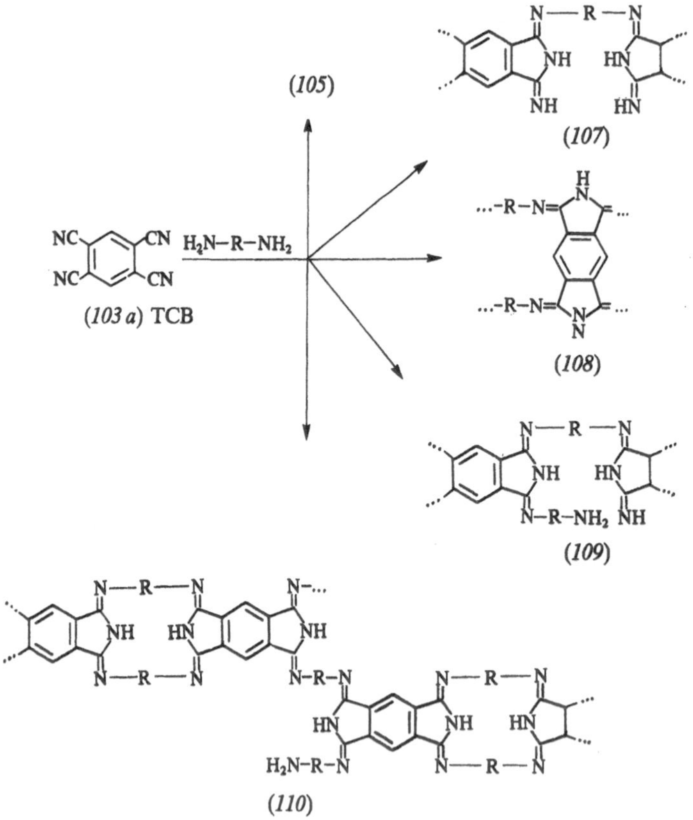

(58)

The reaction of TCB in phenol is very selective in the formation of (105)[259-261]. The best way is heating TCB with diamines in a molar ratio of $1:2$ at 440 K under 860 mbar. The addition of LiCl is increasing the reaction velocity. It is important for a successful condensation that many intermediate products are solved in phenol during polycondensation.

The brown to black coloured mainly amorphous polymers obtained in high yield ($<95\%$) are often soluble in polar solvents. In one case molecular weights of $\sim 19\,000$ were observed by light scattering[259].

Model reactions between 1,2-dicyanobenzene and $(104\,l)$ helped to study the mechanism of polyhemiporphyrazine formation in phenol[259, 261]. The rate determining step is the formation of the reactive intermediate 1-phenoxy-3-imino-isoindolenin from nucleophilic addition of phenol to the nitrile. Afterwards the equilibrium rate with the diamine $(104\,l)$ for the formation of the hemiporphyrazine ring is quicker. Using the weaker basic $(101\,k)$ the second step may be rate determining. Normally, the rate of Hp-formation increased in the order $(103\,d) < (103\,c) < (103\,a)$ when using $(104\,a)$; and $(104\,f) < (104\,l) < (104\,a)$ when starting from $(103\,a)$[261]. The reaction is also very sensitive against the kind of phenolic solvent. Phenols with electron attracting groups lead to other stable products[259, 262]. The polycondensation of TCB with $(104\,g)$ and $(103\,c)$, $(103\,d)$ with $(104\,a)$ in phenol is performed in a homogenous phase. The isolated polymers show better thermomechanic properties than the polymers precipated during polycondensation. The soluble polymers have a flow temperature of 320–570 K under 25 Kp/cm². After manufacturing and then heating higher then 480 K the polymers become insoluble due to network formation[261].

The reaction of different tetranitriles (103) and diamines (104) were carried out in bulk at 470–600 K[263-265]. According to the amount of evolved NH_3 the polycondensation seems to be a two step process[263]. At first with a rapid rate ~ 1 mole NH_3 was evolved (inserting 1 mol (103) and 2 mols (104)) leading to the polymer (109). Slow evolution of NH_3 followed with formation of (105) in high yield. The amount of NH_3 allows to conclude that the polymer contains mainly structure elements (105) and (109) perhaps with small amounts (107) as reactive intermediate (Eq. 59). The disadvantage of the bulk procedure is the insolubility of all polymers showing also the presence of (110).

$$
\begin{array}{l}
\text{TCB} + H_2N-R-NH_2 \longrightarrow (107) \\
\qquad\qquad (104) \\[1em]
\dfrac{+ H_2N-R-NH_2}{-NH_3}\ (109) \xrightarrow{-NH_3} (105)
\end{array}
\tag{59}
$$

Much work was done to study the condensation of TCB $(103\,a)$ and different diamines (104) in alcohols like 2-methoxyethanol or methanol with the corresponding sodium alkoxide as catalyst[260, 263, 266-268]. The disadvantage of this procedure is often polymer precipitation at the beginning of NH_3 evolution. The polycondensation is therefore incomplete when using various diamines; this is shown by $C\equiv N$ absorption in the IR spectra of the polymers[260, 263] and the small amount of liberated NH_3.

The condensation products of 1,2-dicyanobenzene and $(104\,a)$ in alkoxide catalysed reactions contain linear oligomers (111)[267]. Therefore polymers starting from TCB may have also the structure elements (108). By heating TCB and $(104\,h)$ in a molar ratio of

1:1 in methanol/sodium methoxide, a small amount of NH_3 was evolved and a dark coloured polymer mainly with structure elements (107) was isolated[268]. The structure of polymers in the system alcohol/alkoxide seems to be very heterogeneous depending on molar ratio, kind of diamine and solvent.

(111)

In some cases dark brown to black coloured metal chelates (106) of the mentioned dark coloured ligands (105) with the diamine (104e) were synthesized (Eq. 57). Two ways of metalisation were used:
– Introduction of metal ions into the polymer ligand, with $CuCl_2$ in DMF only 50–70%[264, 265, 267].
– Heating TCB, (104a, e) and metal acetylacetonates in a molar ratio of 1:2:1 in bulk[264, 265]. Complete metalization was performed by Cu^{2+}, Ni^{2+} and 70–90% introduction by Fe^{2+}, Mg^{2+}. The main difficulty of this procedure is to exclude polyphthalocyanie formation.

The condensation of 1,2-dicyanobenzene or TCB with tretraamines (in o-position e.g. 3,3',4,4'-tetraaminodiphenyl) or triamines led in the alkoxide catalyzed reaction to other polymers with benzoylenebenzimidazole units[269, 270].

By heating TCB with diamines like (104a) and anthranilic acid in polyphosphoric acid, polymers containing no macroheterocycles were formed[271, 272]. The condensations led to polymers with isoindole and quinazoline rings.

The secondary amino groups of low molecular hemiporphyrazine [from 1,2-dicyanobenzene and (104a)] react with bis-(acid chlorides) in dimethylacetaminde/triethylamine at low temperature to polyamidohexaazacyclanes (112). In (112) the center is blocked against metal introduction[273].

(112)

Mainly thermal[259, 261, 266, 268, 270, 274, 275] and electrical properties[259, 261, 264, 265, 275, 276] were investigated.

In general, all polymers containing different structure elements are stable in air up to 630–770 K when heated at a rate of 4–15°/min. The main advantage is that soluble polymers (Table 17) with lower molecular weight or no network formation could be moulded at 570–720 K at pressures of ~ 10^3 lb/in^2. Quite strong specimens were obtained. After determining ultimate tensile strength, flexural strength and flexural modulus the mechanical properties are comparable with commercial polyimides[266]. Very often the

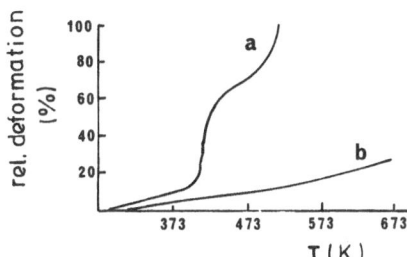

Fig. 12. Thermomechanical properties of polymers from (*103 c*) and (*104 a*) in phenol. (*a*) soluble polymer, (*b*) after heating this polymer to 500 K

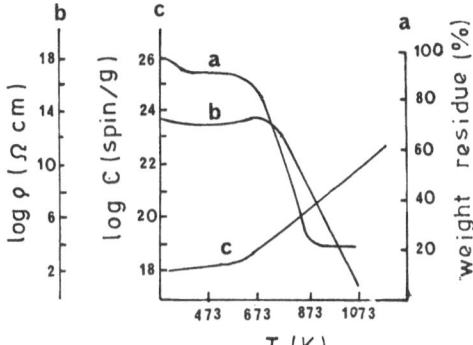

Fig. 13. Dependence of thermal stability (*a*), specific resistance (*b*) and concentration of paramagnetic centers (*c*) on the heat treatment of polymer from (*103 a*) and (*104 a*) prepared in phenol

network formation must occur during manufacturing because the polymer is getting insoluble[259]. Then polymers are resistant against thermomechanical deformation (Fig. 12)[261].

Apparent activation energies and Arrhenius factors were studied for the thermal degration of some polymers and correlated with the evolved gas[274]. In air thermal degradation is of first order. The exothermic breakdown in air at T > 550 K starts with the oxidation of the diamine component in the polymer. For the polymer of TCB and (*104 a*) prepared in phenol a correlation between thermal stability, specific conductivity, and concentration of paramagnetic centers was observed (Fig. 13)[259, 275]. Up to ~ 670 K the polymer remains unchanged (from IR) only with decrease of ordering (from X-ray). At higher temperatures thermal decomposition is correlated with the increase in conductivity and paramagnetic centers. The polymer retained at ~ 1100 K is of charcoal type with perhaps graphite like structure.

The specific conductivity of polymers prepared from (*103 a, b, e, f*) and various diamines ranges between 10^{-8} and to 10^{-15} Ω^{-1} cm^{-1}; [259, 264, 265, 275, 276]. In general, the introduction of metal ions will not enhance the conductivity drastically. In one case 10^{-2} Ω^{-1}cm^{-1} was observed for a polymer from 2,5-diamino-3,4-dicyanothiophene[264].

5.4 Polymeric Tetraaza(14)Annulenes

Monocyclic, bicyclic and polymeric tetraaza(14)annulene ligands and their metal chelates (*113*)–(*117*) were prepared from propynal and amines (Eq. 60)[277, 278].

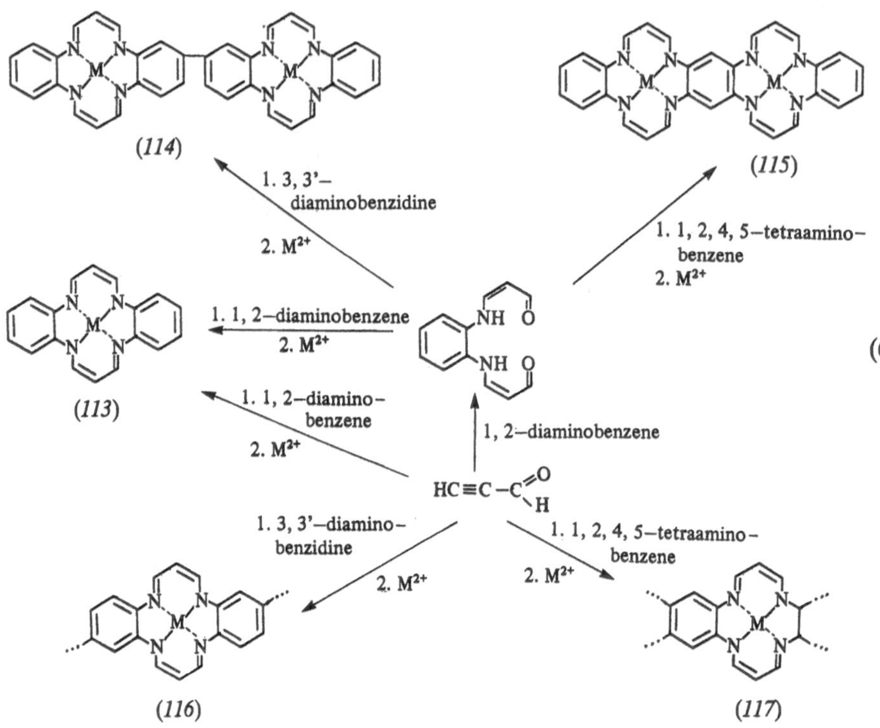

(60)

The structures of all products (113)–(117) were determined by UV/VIS, IR and mass spectra. However, the dark coloured polymers are insoluble in organic solvents. Therefore no molecular weight determination was possible. The thermal decomposition of the polychelates starts under N_2 at ~700 K. At 1070 K weight loss is ~10%[279].

The Co and Cu containing chelates show high *activity as catalyst* for the oxidation of ethylbenzene to acetophenone. A small amount of pyridine is necessary for the activation of the catalyst (Eq. 61)[279].

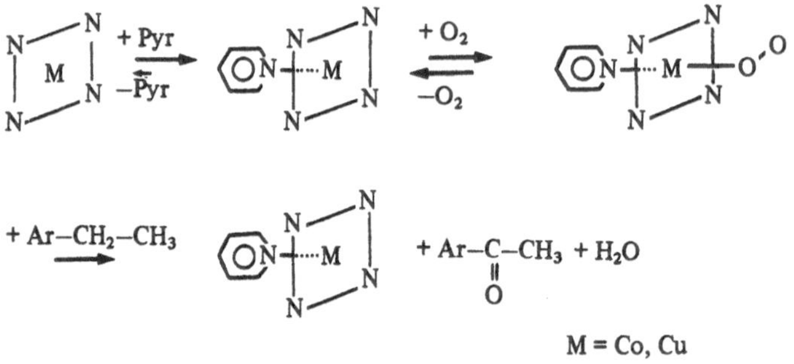

(61)

M = Co, Cu

The activity of the bicyclic chelates are higher than those of polymeric and monocyclic: (115 Co) 60%, (117 Cu) 16%, (113 Co) 5% acetophenone after 6 h at 409 K using 10^{-2} mol% of the catalyst. Ethylbenzene is oxidized stepwise. The high activity of (115) may correlate with the bifunctional contact center of such a process. The lower activity of the polymer may be due to smaller solubility.

For all chelates the *electrical conductivity* was investigated[279]. The low conductivity of $\sigma \sim 10^{-9} - 10^{-12} \, ohm^{-1}cm^{-1}$ may be explained from the X-ray investigation: the delocalisation of π-electrons is concentrated on the 6 membered propan-1,3-diiminato-chelate ring. So intramolecular charge transfer is inhibited.

5.5 Polymeric Dioxime Chelates

When starting from bifunctional bis(1,2-dioximes) in the reaction with metal ions, coordination polymers were prepared by polycondensation.

Bis(1,2-dioximinopropyl)phenyl-alkanes or -ethers suspended in methanol are reacting with Ni(II) to coloured polymers $(118)^{[280]}$. For the pyridine soluble polymer (118) with R = 0, the molecular weight is $\sim 2 \cdot 10^4$. The intractability of the Ni-containing polymers was improved by reacting the –NOH end groups with epichlorhydrin. The obtained 0-epoxy ether group are cured with diamines leading to amine cured epoxy resins. These adhesives were tested in tensile shear tests and showed good properties.

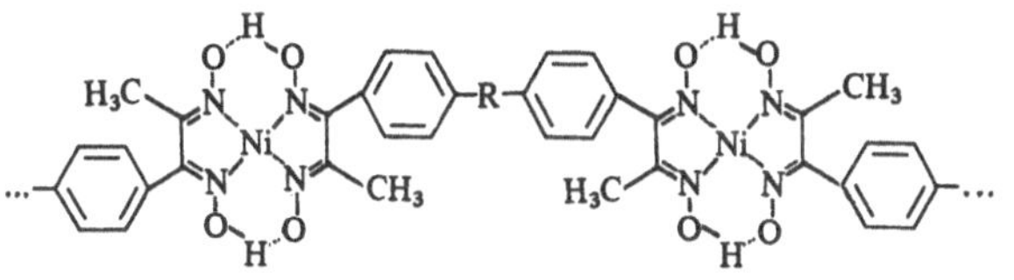

(118)

2,3,6,7-octanetetraonetetraoxime (119) was used as monomer in the reaction with Fe(II), Co(II), Ni(II), Cu(II), Pd(II), Rh(III)[281-283]. Yellow to brown coloured polychelates, containing mainly the structure element (120), were isolated in dioxane (Eq. 62). A pyridine soluble fraction showed a molecular weight of 2500.

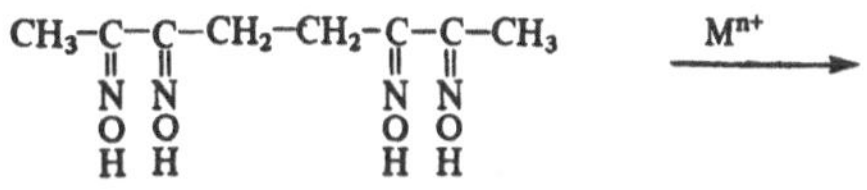

(119) (62)

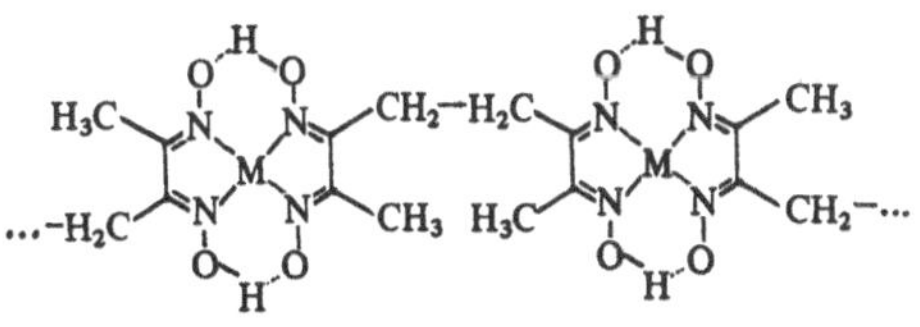

(120)

The interesting properties of (120) are:
– reversible binding of O_2 and CO
–hydrogenation reactions

The *binding of small molecules* like O_2 and CO was examined in bulk or solvents[282]. As to be seen from Table 18 the Fe(II)-, Co(II)- and Cu(I)-chelates are able to bind oxygen depending on the solvent. CO uptake is possible with the Fe(II)-, Pd(II) and

Table 18. Oxygen uptake of the polymeric oximes $(120)^{282)}$, $(122)^{284)}$ at 293 K

Compound	Mole ratio of absorbed O_2/metal[a]			
	Without solvent	in CH_3OH	in H_2O	in benzene
(120) Fe(II)	small	0,22 (21)	–	0,27 (34)
(120) Co(II)	small	0,11 (63)	–	–
(120) Ni(II)	0	0	0	0
(120) Cu(I)	small	0,18 (41)	0,18 (14)	–
(122) Cu(I)	–	0,14 (~90)	0,14 (~90)	–

[a] values in parenthesis: reversibility in %

Table 19. Hydrogenation of alkenes and alkynes by Rh(III) and Pd(II) chelates of (120). Reaction at 293 K at 1 atm H_2 with 40 mg (120), ~1,2 mmol substrate in 5 ml ethanol

Metal in (120)	Reaction time/h	Substrate	Product	Yield/ %
Rh	0,5	PhC≡CPh	$PhCH_2CH_2Ph$	100
Pd	24	PhC≡CPh	$PhCH_2CH_2Ph$	100
Rh	24	2-heptene	heptane	20
Pd	18	2-heptene	heptane	95
Rh	72	cyclohexene	cyclohexane	25
Pd	24	cyclohexene	cyclohexane	100
Rh	120	1,3-Cyclooctadiene	–	0
Pd	24	1,3-Cyclooctadiene	cyclooctane	100
Rh	24	D-limonene	menthene	97
Pd	8	D-limonene	menthene	52
			menthane	42

Cu(I)-chelates. The reversibility of O_2 uptake is decreasing continously from cycle to cycle due to oxidation of the metal.

Rh(III)- and Pd(II)-chelates of (120) are active catalysts for the hydrogenation of alkenes and alkynes under mild conditions (Table 19)[283)].

Normally, the Rh(III) chelate is active for the hydrogenation of
– terminal double and triple bonds,
– inner double and triple bonds activated by conjugation with an aromatic ring.
The Pd(II)-chelate catalyses the hydrogenation of
– inner triple bonds activated by an aromatic ring,
– inner and terminal alkenes,
– aromatic ketones, aromatic nitro compounds, reductive halogenation of aryl-halogenides.

It is assumed that the Rh(I) and Pd(0) are the active centers in the hydrogenation. Pd-chelates with higher metal content in the polymer showed a higher activity.

Polyoximes (121) prepared by copolymerisation of ethylene and CO followed by reaction with hydroxylamine were used as starting material. Reaction with different metal ions like Cu(II) and Co(II) in a methanol/dioxane mixture leads to coloured insoluble chelates such as (122) (Eq. 63)[284]. After evaluating the IR spectra it is assumed that also other structure elements than normal bis(glyoximato)Cu(II) are present.

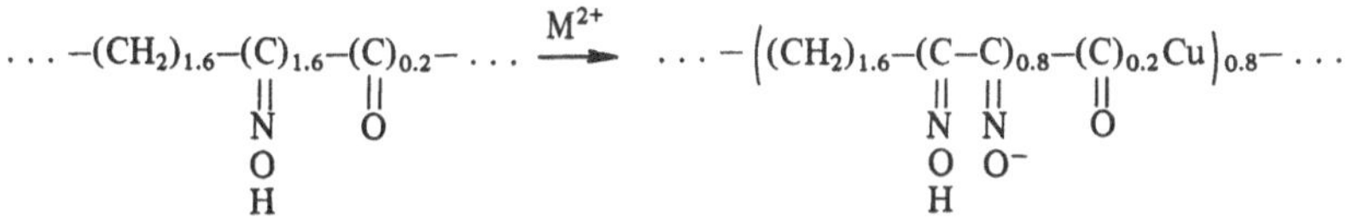

$$\ldots -(CH_2)_{1.6}-(C)_{1.6}-(C)_{0.2}- \ldots \xrightarrow{M^{2+}} \ldots -\left[(CH_2)_{1.6}-(C-C)_{0.8}-(C)_{0.2}Cu\right]_{0.8}- \ldots$$

(63)

(121)   (122)

Good reversible binding of O_2 and CO was observed for Cu(I)-chelate (Table 18)[284]. The Cu(I)- and Cu(II)-chelates (122) show a high activity in the Michael-type addition of alcohols to acrolein[284]. Up to 66 mol β-alkoxypropionaldehyde per mol metal center were obtained; the yield decreased with lower Cu content in (122) Acrylonitrile is polymerized in the presence of (122)-Cu(II) under H_2 pressure. The Co containing complex is able to polymerize styrene and to catalyze Michael additions. For a Pd-complex CO binding and afterwards catalytic hydrogenation of alkenes are reported[284].

5.6 Polymeric N₂O₂-Chelates

As mentioned in the introduction, the synthesis of low molecular N_2O_2-chelates as Co(salen) (6) is rather easy when starting from salicylaldehyde, diamines and a metal salt. For the polymer synthesis only bifunctional o-hydroxyaldehydes must be used. The limiting factor for practical use of polymers may be the price of the bifunctional monomer. The difficulty in preparation is to get a structurally pure polymer and to reach quantitative metallization.

5.6.1 Synthesis

The following bifunctional o-hydroxyaldehydes (123) and diamines (124) were used to prepare at first the polymer ligand (125), (126). The introduction of metal ions like Fe(II, III), Co(II, III), Ni(II), Cu(II), Zn(II), Cr(II, III), Rh(III), Pd(II), UO₂(II), Ti(IV), V(IV), In(III) led to the polychelates (125 M), (126 M). If the degree of polycondensation is not too high, the yellow polymer ligands are soluble in polar solvents. The intensively coloured polychelates are insoluble.

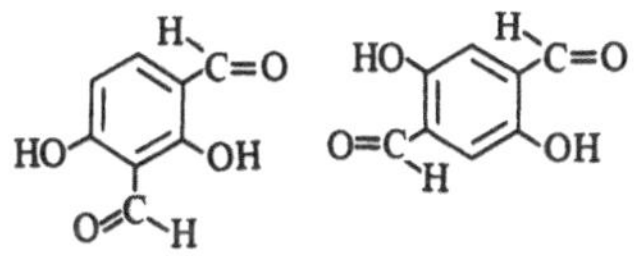

(123 a)   (123 b)

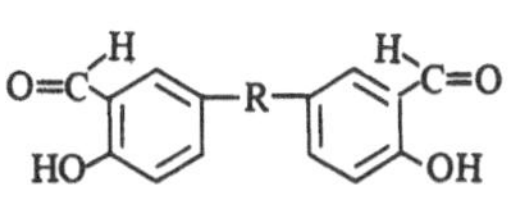

(123 c) R = –CH₂–
(123 d) R = –SO₂–

$$H_2N-(CH_2)_n-NH_2 \qquad H_2N-\overset{\overset{\displaystyle CH_3}{|}}{CH}-CH_2-NH_2 \qquad H_2N-(CH_2)_2-NH-(CH_2)_2-NH-(CH_2)_2-NH_2$$

(124a) n = 1 – 12   (124b)   (124c)

$$H_2N-(CH_2)_3-\overset{\overset{\displaystyle R}{|}}{N}-(CH_2)_3-NH_2$$

(124d)   (124e)

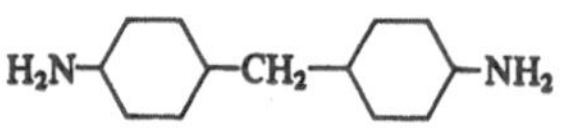

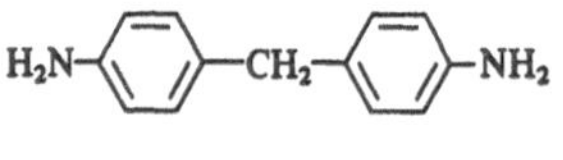

(124f) 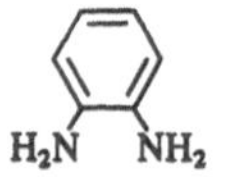   (124g) 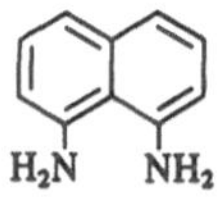   (124h) 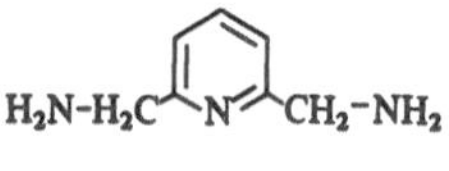   (124i)

When starting from (123 a, b), ladder type polychelates (125 M) from (123 c, d) single bond connected polychelates (126 M) were isolated. Three routes of preparation were examined (Table 20, Eq. 64)[286–295]:

Route a) Reaction of (123) with diamines to prepare the yellow to brown coloured polymer ligands (125), (126). Afterwards insertion of metal ions in a separate step.

Route b) One step reaction by heating all components together.

Route c) At first preparation of an O_4-chelate polymer (127) from (123) and metal salts and in a separate step addition of the diamine.

If the polymer ligands (125), (126) are soluble in polar solvents like DMAc or DMF route a) is the best decision. By adding metal salts, the polychelates are precipitated. Also structure investigation is easier with polymer ligands than with insoluble polychelates. Route b) may be choosen if the polymer ligand is insoluble because quantitative introduction of metal ion using route a) may be without success.

Table 20. Preparation of polymeric N_2O_2-chelates

Hydroxy-aldehyde	Diamine	Route	Ligand preparation	Metal introduction	Structure polychelates	Ref.
(123 c, d)	(124 g)	a	THF/CH₃COOH or emulsion condensation	DMF	(126 M)	285
(123 a)	(124 a) n = 2	a	Ethanol	Ethanol	(125 M)	286
(123 c, d)	(124 c, i)	a	Ethanol	DMF	(126 M)	287
(123 b)	(124 a) n = 1–4; (124 g, h)ª	a b c	CH₃COOH CH₃COOH dioxan	DMAc CH₃COOH DMAc	(125 M)	288
(123 c)	(124 a) n = 2–8, 10, 12 (124 b); (124 d) R=CH₃; a (124 e, f, j, h)		DMF	DMF	(126 M)	289–293
(123 c)	(124 b)	a	DMF	DMF	(126 M)	294
(123 c)	(124 f)	a	THF/CH₃COOH	DMF	(125 M)	295

ª Also different hydroxydiamines used

When using route c) some side reactions may occur. In the reaction of bis-(salicylaldehydo)metal(II)chelates (*128*) (M = Be^{2+}, Zn^{2+}, Cd^{2+}) with diamines like (*124g*) no low molecular N_2O_2-chelate (*129*) was obtained in DMF (Eq. 65)[296]. So using (*123*) in route c) crosslinking may occur.

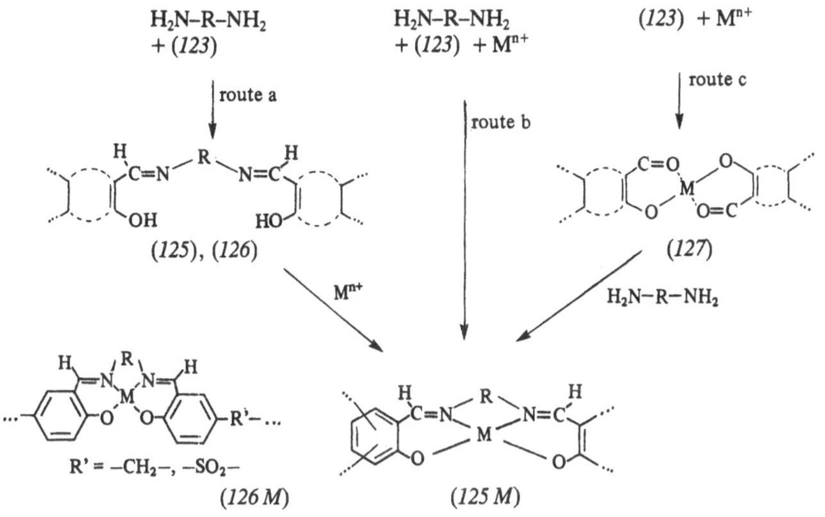

(64)

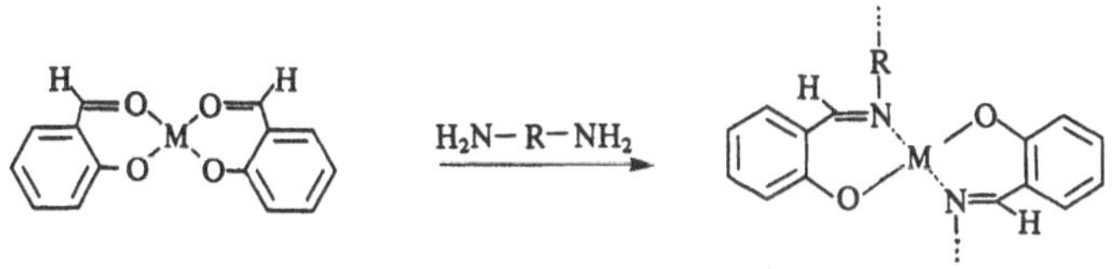

(65)

(*128*) (*129*)

N_2O_2-chelates (*131*), (*133*) with an open bridge were prepared by reaction of bifunctional o-hydroxyaldehyde derivatives (*130*), (*132*) with metal salts in DMF or ethanol (Eq. 66, 67)[288, 297, 298].

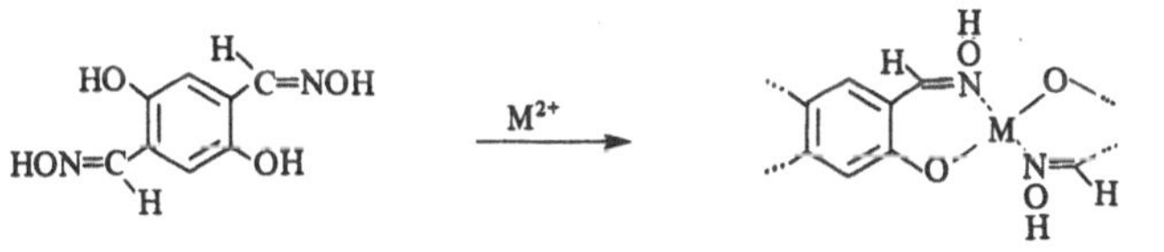

(66)

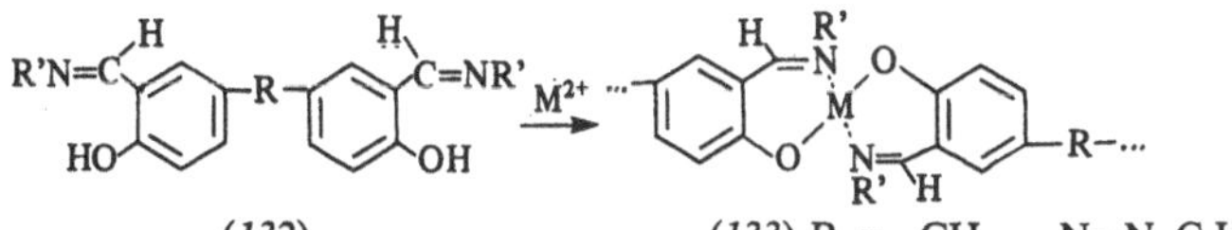

(*130*) (*131*) M = Ni(II), Cu(II)

(67)

(*132*) (*133*) R = $-CH_2-$, $-N{=}N-C_6H_4-N{=}N-$
R′ = $-C_6H_5$, $-OH$
M = Mn(II), Zn(II), Co(II), Ni(II), Cu(II), Cr(III)

Finally, a brown coloured insoluble polychelate (*135*) of acacen-type was prepared from (*134*) and (*124g*) in the presence of Ni(II) (Eq. 68)[299].

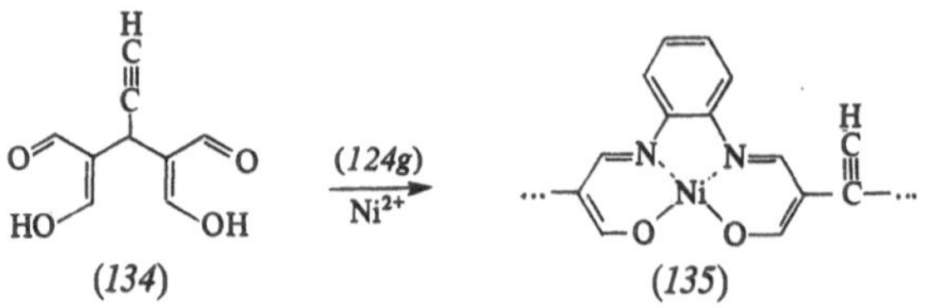

$$\tag{68}$$

Polymeric N_2O_2-chelates may be obtained by polyreaction of a bifunctional low molecular N_2O_2-chelate instead of constructing the chelate system during polycondensation. But only few results show a new way for the future. Bifunctional low molecular chelates (136) solved in NaOH were condensed with bifunctional aromatic acid chlorides solved in CH_2Cl_2 by interfacial polycondensation[300]. Insoluble polychelates (137) were obtained (Eq. 69).

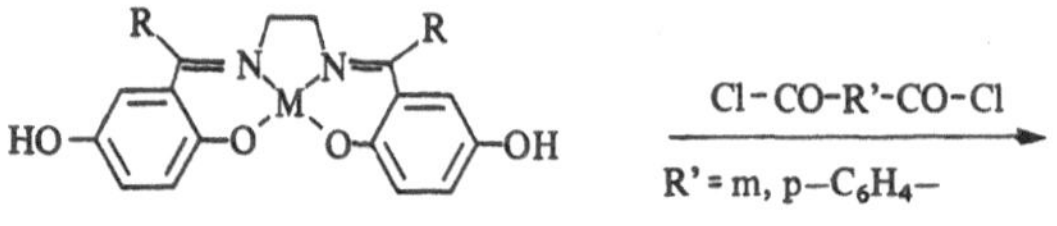

(136) M = Cu(II), Ni(II)
R = –H, –CH$_3$, –C$_6$H$_5$

$$\tag{69}$$

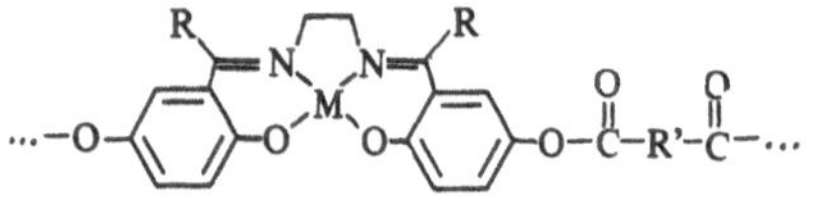

(137)

The mentioned polymer ligands and polychelates are mainly characterized by IR spectra and elementary analysis. Also results about magnetic measurements and visible reflectance spectra are reported[289, 293, 295–298, 300]. For the polymers (126 M) from (123 c) and (124 f) the following arrangement of the ligand atoms around the metal atoms is assumed[295]: planar structure (M = Cu(II)) tetrahedral structure (Co(II)) distorted octahedral symmetry (Ni(II), Mn(II)). The following structures are suggested for chelates (129): Ni(II), Cu(II) square planar; Co(II) tetrahedral[296]. Chelates (133) (R' = OH) are mainly existing in a octahedral arrangement; (133) with R' = C_6H_5, octahedral for Ni(II) and tetrahedral for Co(II) is described[138].

5.6.2 Properties

Main research interest until now was getting informations about *thermal stability* of polymer N_2O_2-chelates[285, 287, 290, 293, 295–298, 300]. In air or nitrogen the decomposition of the polychelates mainly occurs between 520 and 720 K. The thermal stability of the polychelates (126 M) depends on the following parameters.
– type of bissalicylaldehyde[285, 287]: $CH_2 < SO_2$
– metal atom[293, 295]: $Co^{2+} < Zn^{2+} < Cr^{2+} < Fe^{2+} < Pd^{2+} < Cu^{2+} < Be^{2+} < Ni^{2+}$;
    $Ti^{4+} < Fe^{3+} < Co^{3+} < Rh^{3+} < U^{6+} < Cr^{3+} < In^{3+}$.
– diamines[296]: (121 a) with $(CH_2)_3 > (CH_2)_2 > (CH_2)_{4-11}$

In (126) from (123c) and (124f) the polymer ligand has a higher thermal stability of ~ 770 K than the polychelates of 520–620 K (with Zn > Ni > Mn > Co > Cu)[295]. Some new future aspects of properties were investigated in the last years[289, 291]:

The *adsorption of solvent molecules* by chelates (126 M) from (123c) and (124e) with bivalent metal cations of high coordination number, e.g. Cr^{2+} was studied. So molecules such as pyridine, CCl_4, benzene, CH_3OH, n-pentane with a critical diameter of < 7 Å are adsorbed. The capacity is higher than using various molecular sieves (except for water).

These polychelates are used as *stationary phase* in molecular sieve GC to separate gas mixtures of rare gases.

Polymeric ligands of the type (126) from (123c) and (124b) were investigated for the selective *metal binding* from DMF solutions[294]. The following sequence of complexation was found:

Ni > Cu > Fe > Zn .

Semiconducting properties were investigated[288, 295]. Conductivities are in the order of $10^{-10} - 10^{-14} \, \Omega^{-1}cm^{-1}$ due to comparatively small intra- and intermolecular charge transfer of polymer ligands and polychelates.

6 Future Aspects

In general, many synthetic procedures are described for planar metal chelates in order to construct active compounds for practical use. Most important is the reproducibility of composition and the structural uniformity. Only in this way statements about important properties are applicable for practical use. The properties are located in a broad field of saving natural resources and conducting or transforming energy.

Soluble, swellable and macroporous chelate polymers with coordinative and covalent bonds (Chaps. 2, 3) may be prepared. Structure investigations are in most cases possible with conventional methods. The advantage of coordinative binding is the ease of preparation. But on the other side such a bond is not so strong when compared with a covalent one. So the application must decide between coordinative or covalent bond. Reversible binding of small molecules, catalysis and photoredox reactions may be important. Cheaper, easier to prepare and more stable phthalocyanines, oximes and Schiff'base chelates will find higher practical interest then porphyrins.

The disadvantage of covalent binding of the chelate in a narrow arrangement in the staple (Chap. 4.1) is the comparatively low solubility and infusibility. Polymers containing N_4-chelates with coordinative or combined covalent coordinative bond may have the advantage of better solubility and sublimation for the preparation of thin films. Electrical properties are important because conductivity ranges from the semiconductor to nearly the metal region. But much work must be spent to examine other kinds of electrical properties such as photoconductivity, arrangements for photovoltaic and photogalvanic elements.

N_4-chelates covalent bond in another polymer (Chap. 4.2) are interesting either as dye-stuffs with covalent bond or thermally stable compounds.

Polymeric phthalocyanines (Chap. 5.2) include a great variety of properties. The construction of electrical devices or catalysts for special use is most hopeful. But all these applications depend on the reproducibility of well defined structural uniform polymers. Preparative work must help to standardize synthetic procedures and to investigate structures. The well reproducible in situ synthesis of thin layers of pure polymeric phthalocyanines from the gas phase opens a way for electrocatalysis and visible light energy changing devices. Another new kind of preparation goes via prepolymers which may be converted to mechanically and thermally stable, infusible polymers.

The possibility of getting structurally pure polymeric hemiporphyrazines is very small. Intensive investigation of model reactions and more detailed structure determination by analytical methods of soluble polymer ligands may help. Manufacturing of fusible polymer ligands leads to products of high thermal and chemical resistance. Only the price of the tetranitrile is the limiting factor.

Metal containing polyoximes (Chap. 5.6) are interesting compounds for binding small molecules (O_2, CO) and activating H_2 for catalytic reactions. The reversibility for O_2-binding in water seems to be higher than with polymer bound porphyrins. Structurally pure polyoxime chelates are difficult to obtain since other structural elements are often present.

Possibilities for preparing polymeric N_2O_2-chelates (Chap. 5.6) and their ligands are known. Some work must be done in structure investigation by molecular weight and its distribution. Interesting properties are their use as adsorbents. The research about binding small molecules like O_2 and application as catalyst is yet missing.

The comparison of main properties leads to the following statement:

O_2-binding ability: Pc < oximes < schiffbase < P   (polymer bond)
O_2-reversibility (Scheme 1): P < oximes < schiffbase   (polymer bond)
low temp. O_2 catalysis: P < schiffbase, Pc   (polymer bond)
high temp. catalysis: oximes, schiffbase < Taa < Pc   (polymeric)
thermal stability: Pc, schiffbase, Hp   (polymeric)
electron transport: schiffbase, Hp, Taa < Pc   (polymeric)
fuel cell: Hp, schiffbase < Taa < P < Pc   (polymeric)
visible light energy conversion: Hp, schiffbase < Taa < Pc, P

7 References

1. Wöhrle, D.: Adv. Polym. Sci., *10*, 35 (1972)
2. Smith, K. M. (ed.): Porphyrins and metalloporphyrins, Amsterdam, Elsevier Scientific Publishing Company 1975
3. Fuhrhop, J.-H.: Angew. Chem. *88*, 704 (1976)
4. Buchler, J. W.: Angew. Chem. *90*, 425 (1978)
5. Berezin, B. D.: Coordination Compounds of Porphyrins and Phthalocyanines, N. Y., John Wiley and Sons 1981
6. Moser, F. H., Thomas, A. L.: Phthalocyanine compounds, New York, Reinhold Publishing Corporation 1963
7. Lever, A. B. P.: Advan. Inorg. Chem. Radiochem. *7*, 27 (1965)
8. Kasuga, K., Tsutsui, M.: Coord. Chem. Rev. *32*, 67 (1980)
9. Bamfield, P., Mack, P. A.: J. Chem. Soc., *1968* (1961)
10. Hecht, J. A., Luger, P.: Acta Cryst. B *30*, 2843 (1974)

11. Hiller, H., Dimroth, P., Pfitzner, H.: Justus Liebigs Ann. Chem. *717*, 137 (1968)
12. Schrauzer, G. N.: Angew. Chem. *88*, 465 (1976)
13. Jones, R. D., Summerville, D. A., Basolo, F. Chem. Rev. *79*, 139 (1979)
14. Basolo, F., Hoffmann, B. M., Ibers, J. A.: Acc. Chem. Res. *8*, 384 (1975)
15. McCleverty, J. A.: Chem. Rev. *79*, 53 (1979)
16. Tsuchida, E., Nishide, H.: Adv. Polym. Sci. *24*, 1 (1977)
17. Blauer, G.: Acta chem. Scand. *17*, 8 (1963); Biochim. Biophys. Acta *133*, 206 (1976); Arch. Biochem. Biophys. *121*, 587 (1967)
18. Yamamoto, S., Nozawa, T., Hatano, M.: Polymer *15*, 330 (1974)
19. Tsuchida, E., Honda, K., Hasegawa, E.: Biochim. Biophys. Acta *393*, 483 (1975)
20. Tsuchida, E., Hasegawa, E., Honda, K.: Biochim. Biophys. Acta *427*, 520 (1976)
21. Wang, J. H., Bringar, W. S.: Proc. Natl. Acad. Sci., U.S. *45*, 958 (1960)
22. Tsuchida, E., Hasegawa, E., Honda, K.: Biochim. Biophys. Res. Commun. *67*, 864 (1975)
23. Tohjo, M., Shibata, K.: Arch. Biochem. Biophys. *103*, 401 (1963)
24. Tsuchida, E., Hasegawa, E., Honda, K.: J. Polym. Sci., Polym. Chem. Ed. *13*, 1747 (1975)
25. Tsuchida, E., Honda, K., Sata, H.: Biopolymers *13*, 2147 (1974)
26. Tsuchida, E. et al.: Chem. Letters *1975*, 761
27. Honda, K., Hata, S., Tsuchida, E.: Bull. Chem. Soc. Japan *49*, 868 (1976)
28. Tsuchida, E., Honda, K., Sata, H.: Makromol. Chem. *176*, 2251 (1975)
29. Tsuchida, E., Honda, K., Sata, H.: Inorg. Chem. *15*, 352 (1976)
30. Tsuchida, E., Hasegawa, E., Ohno, H.: J. Polym. Sci., Polym. Chem. Ed. *15*, 561 (1977)
31. Nishide, H., Mihayashi, K., Tsuchida, E.: Biopolymers *18*, 739 (1979)
32. Rubin, L. B. et al.: Dokl. Akad. Nauk SSSR *258*, 1472 (1981)
33. Hatano, M.: Kagaku to Kogyo *18*, 926 (1965)
34. Nishide, H., Ohno, H., Tsuchida, E.: Makromol. Chem. Rapid. Commun. *2*, 55 (1981)
35. Nishide, H., Ohno, H., Tsuchida, E.: Kobunshi Robunshu *37*, 641 (1980)
36. Scherer, W. et al.: Eur. J. Biochem. *13*, 77 (1970)
37. Tsuchida, E. et al.: J. inorg. nucl. Chem. *40*, 1241 (1978)
38. Allcock, H. R.: J. Am. Chem. Soc. *101*, 606 (1979)
39. Kühn, M., Benes, M.: Z. Chem. *20*, 351 (1980)
40. Kühn, M., Ristan, O., Coupek, J.: Z. Chem. *21*, 231 (1981)
41. Fuhrhop, J.-H. et al.: Makromol. Chem *178*, 1621 (1977)
42. Chang, C. K., Traylor, T. G.: Proc. Nat. Acad. Sci., USA, *70*, 2647 (1973)
43. Wang, J. H.: Acc. Chem. Res. *3*, 3168 (1970) and references therein
44. Nishide, H., Tsuchida, E. et al.: Biopolymers *17*, 191 (1978)
45. Tsuchida, E., Shigehara, K., Miyamoto, K.: J. Polym. Sci., Polym. Chem. Ed. *14*, 911 (1976)
46. Collmann, J. P., Reed, C. A.: J. Am. Chem. Soc. *95*, 2048 (1973)
47. Basolo, F. et al.: J. Am. Chem. Soc. *97*, 5125 (1975)
48. Fuhrhop, J.-H. et al.: Angew. Chem. *88*, 616 (1976)
49. Tsuchida, E.: J. Macromol. Sci., Chem. *A 13*, 545 (1979)
50. Shigehara, K., Tsuchida, E. et al.: Macromolecules *14*, 1153 (1981)
51. Tsuchida, E., Hasegawa, E.: Biopolymers *16*, 845 (1977)
52. Nishide, H., Mihayashi, K., Tsuchida, E.: Biochim. Biophys. Acta *498*, 208 (1977)
53. Kokufuta, E., Watanabe, N., Nakamura, I.: J. Appl. Polym. Sci. *26*, 2601 (1981)
54. Iwabuchi, S. et al.: J. Polym. Sci., Polym. Lett. Ed. *19*, 193 (1981)
55. Iwabuchi, S. et al.: Res. Rep. Fac. Eng. Chiba Univ. *33*, 25 (1981)
56. Politis, T. G., Drickamer, H. G.: J. Chem. Phys. *74*, 263 (1981)
57. Kobayashi, H. et al.: J. Phys. Chem. *86*, 114 (1982)
58. Hata, S., Tsuchida, E.: Polym. Prepr; Am. Chem. Soc., Div. Polym. Chem. *20*, 514 (1979)
59. Zwart, J. et al.: J. Mol. Catal., *3*, 151 (1977)
60. Schutten, J. H., Zwart, J.: J. Mol. Catal. *5*, 109 (1979)
61. Schutten, J. H., Piet, P., German, A. L.: Makromol. Chem. *180*, 2341 (1979)
62. Przywarska-Boniecka, H., Trynda, L., Antonini, E.: Eur. J. Biochem. *52*, 567 (1975)
63. Yamada, A., Tsuchida, E. et al.: Makromol. Chem. *181*, 1823 (1980)
64. Tsuchida, E. et al.: J. Polym. Sci., Polym. Chem. Ed. *17*, 807 (1979)
65 a. Lautsch, W., Broser, W., Gnichtel, H.: J. Polym. Sci. *17*, 479 (1955) and lit. cited there.
65 b. Lautsch, W. et al.: Kolloid Zeitschrift *125*, 72 (1952)

66. Kühn, M. et al.: J. Polym. Sci., 47, 69 (1974)
67. Yamakita, H., Hayakawa, K.: J. Polym. Sci., Poly. Lett. Ed. 18, 529 (1980)
68. Kamogawa, M.: J. Polym. Sci., Polym. Chem. Ed. 12, 2317 (1974); J. Polym. Sci., Polym. Lett. 10, 711 (1972)
69. Kamogawa, H., Inoue, H., Nanasawa, M.: J. Polym. Sci., Polym. Chem. Ed. 18, 2209 (1980)
70. Shigehara, K., Tsuchida, E.: Polym. Prepr., Am. Chem. Soc., Div. Polym. Chem. 20, 1057 (1979)
71. Rollmann, L. D.: J. Amer. Chem. Soc. 97, 2132 (1973)
72. King, R. B., Sweet, E. M.: J. Org. Chem. 44, 386 (1979)
73 a. Lennox, J. C., Murray, R. W.: J. Am. Chem. Soc. 100, 3710 (1978);
73 b. Rocklin, R. D., Murray, R. W.: J. Phys. Chem. 85, 2104 (1981)
74. Bettelheim, A., Chan, R. J. H., Kuwana, T.: J. Electroanal. Chem. Interfacial Electrochem. 110, 93 (1980)
75. Nishide, H., Shinohara, K., Tsuchida, E.: J. Polym. Sci., Polym. Chem. Ed. 19, 1109 (1981)
76. Nishide, H., Kato, M., Tsuchida, E.: Eur. Polym. J. 17, 579 (1981)
77. Kokufuta, E., Watanabe, H., Nakamura, I.: Polym. Bull. 4, 603 (1981)
78. Tsuchida, E. et al.: Eur. Polym. J. 14, 123 (1978)
79. Tsuchida, E., Hasegawa, E., Kanayama, T.: Macromolecules 11, 947 (1978)
80. Hasegawa, E., Kanayama, T., Tsuchida, E.: J. Polym. Sci., Polym. Chem. Ed. 15, 3039 (1977)
81. Hasegawa, E., Kanayama, T., Tsuchida, E.: Biopolymers 17, 651 (1978)
82. Nishide, H., Mihayashi, K., Tsuchida, E.: Biopolymers 18, 739 (1978)
83. Bayer, E., Holzbach, G.: Angew. Chem. Int. Ed. Engl. 16, 117 1977)
84. Scharf, H.-D. et al.: Angew. Chem. 91, 696 (1979)
85. Manassen, J.: J. Cat. 18, 38 (1970)
86. Chen, M. J., Feder, H. M.: J. Cat. 55, 105 (1978)
87. Winslow, E. C., Gershman, N. E.: J. Polym. Sci. A, 1, 2383 (1963)
88. Wöhrle, D., Koßmehl, G., Manecke, G.: Makromol. Chem. 154, 111 (1972)
89. Maas, T. A., Kuijer, M., Zwart, J.: J. Chem. Soc. Chem. Commun. 1976, 86
90. Zwart, J., van Wolput, J. H.: J. Mol. Catalysis 5, 235 (1979)
91. Shirai, H. et al.: J. Polymer Sci. Poly. Lett. Ed. 17, 661 (1979); Makromol. Chem. 181, 575 (1980)
92. Schutten, J. H. et al.: Angew. Makromol. Chem. 89, 201 (1980)
93. Gebler, M.: J. inorg. nucl. Chem. 43, 2759 (1981)
94. Kálalová, E., Kátal, J., Svec, F.: Angew. Makromol. Chem. 54, 141 (1976)
95. Svec, E.: Acta Polym. 31, 68 (1980)
96. Bied-Charreton, C., Idoux, J. P., Gaudemar, A.: Nouveau J. de Chimie 2, 303 (1979)
97. Drago, R. S. et al.: J. Am. Chem. Soc. 102, 1033 (1980)
98. Wöhrle, D.: Polymer Bull. 3, 227 (1980)
99. Bohlen, H., Martens, B., Wöhrle, D.: Makromol. Chem. Rapid Commun. 1, 753 (1980); Makromol. Chem. in preparation
100. Aeissen, H., Wöhrle, D.: Makromol. Chem. 182, 2161 (1981)
101. Wöhrle, D. et al.: Makromol. Chem. in preparation; Preprints IUPAC Straßburg 1, 222–225 (1981)
102. Nishinaga, A., Tomita, H.: J. Mol. Cat. 7, 179 (1980)
103. Tsuchida, E., Kurimura, Y.: J. Polym. Sci., Polym. Chem. Ed., 16, 2453 (1978)
104. Joyner, R. D., Kenney, M. E.: Inorg. Chem. 1, 717 (1962)
105. Kroenke, W. J. et al.: Inorg. Chem. 2, 1064 (1963)
106. Esposito, J. M., Kenney, M. E., Sutton, L. E.: Inorg. Chem. 6, 1116 (1967)
107. Sutton, L. E., Kenney, M. E.: Inorg. Chem. 6, 1869 (1967)
108. Meyer, G., Wöhrle, D.: Makromol. Chem. 175, 715 (1974)
109. Hartmann, M., Meyer, G., Wöhrle, D.: Makromol. Chem. 176, 831 (1975)
110. Meyer, G., Hartmann, M., Wöhrle, D.: Makromol. Chem. 176, 1919 (1975)
111. Meyer, G., Wöhrle, D.: Z. Naturforsch. 32 b, 723 (1977)
112. Berezin, B. D., Akopov, A. S., Lapshina, O. B.: Vysokomol. Soedin. A 16, 450 (1974)
113. Berezin, B. D., Akopov, A. S.: Vysokomol. Soedin. A 16, 2334 (1974)
114. Petelenz, P., Zgierski, M. Z.: Mol. Phys. 25, 237 (1973); Hush, N. S., Woolsley, I. S.: Mol. Phys. 21, 465 (1971)

115. Owen, J. E., Kenney, M. E.: Inorg. Chem. *1*, 334 (1962)
116. Final report Aerospace research center, Contract Nobs 92081 (1965)
117. Mitulla, K., Hanack, M.: Z. Naturforsch. *35 b*, 1111 (1980)
118. Hanack, M., Mitulla, K., Schneider, O.: Chemica Scripta *17*, 139 (1981)
119. Hanack, M. et al.: J. Organomet. Chem. *204*, 315 (1981)
120. Schneider, O., Hanack, M.: Angew. Chem. *92*, 391 (1980)
121. Schneider, O., Hanack, M.: Angew. Chem. *94*, 68 (1982)
122. Fuhrhop, J.-H. et al.: Angew. Chem. *92*, 321 (1980)
123. Carraher, C. E., Torre, L. P.: Org. Coat. Plast. Chem. *45*, 252 (1981)
124. Reed, C. A., Scheidt, W. R. et al.: J. Am. Chem. Soc. *100*, 3232 (1978)
125. Kuznesof, P. M. et al.: J.C.S. Chem. Comm. *1980*, 121
126. Linsky, J. P. et al.: Inorg. Chem. *19*, 3131 (1980)
127. Nohr, R. S. et al.: J. Am. Chem. Soc. *103*, 4371 (1981)
128. Kuznesof, P. M. et al.: J. Macromol. Sci.-Chem. *A 16*, 299 (1981)
129. Tsutsui, M.: Gov. Rep. Announce Index (U.S.) *81*, 3348 (1981); Chem. Abstr. *96*, 14553 b
130. Meyer, G., Wöhrle, D.: Materials Science *7*, 265 (1981)
131. Marks, T. J.: J. Am. Chem. Soc. *101*, 7071 (1979)
132. Hanack, M., Seelig, F. F., Strähle, J.: Z. Naturforsch. *A 34*, 983 (1979)
133. Seelig, F. F.: Z. Naturforsch. *A 34*, 986 (1979)
134. Meyer, G., Plieninger, P., Wöhrle, D.: Angew. Makromol. Chem. *72*, 173 (1978)
135. Delman, A. D., Kelly, J. J., Simms, B. B.: J. Polym. Sci. A-1, *8*, 111 (1970)
136. Davison, J. B., Wynne, K. J.: Macromolecules *11*, 186 (1978)
137. Maltzan, B. von: Liebigs Ann. Chem. *1978*, 238
138. Maltzan, B. von: Liebigs Ann. Chem. *1980*, 1082
139. Anton, J. A., Kwong, J., Loach, P. A.: J. Heterocyclic Chem. *13*, 717 (1976)
140. Berlin, A. A., Sherle, A. J.: Inorg. Macromol. Rev. 1, 235 (1971)
141. Wöhrle, D.: Makromol. Chem. *175*, 1751 (1974)
142. Ohsaku, M., Murata, H., Imamura, A.: Europ. Polym. J. *17*, 327 (1981)
143 a. Wöhrle, D.: Makromol. Chem. *160*, 83 (1972);
143 b. – *160*, 99 (1972);
143 c. – *161*, 121 (1972)
144 a. Helling, G., Wöhrle, D.: Makromol. Chem. *179*, 87 (1978);
144 b. – *179*, 101 (1978)
145. Minke, R., Freireich, S., Zilkha, A.: Israel J. Chem. *13*, 212 (1975)
146. Freireich, S., Gertner, D., Zilkha, A.: Europ. Polym. J. *9*, 411 (1973)
147. Biedermann, H.-G., Wichmann, K.: Chemiker-Z. *98*, 161 (1974)
148. Zhubanov, B. A. et al.: Dokl. Akad. Nauk *246*, 365 (1979)
149 a. Subramanian, R. V. et al.: Polymer Bull. *1*, 421 (1979)
149 b. – Adv. Polym. Sci. *33*, 33 (1979)
150. Migahed, M. D., Beckey, H. D.: Kolloid-Z., U.Z. Polymere *246*, 679 (1971)
151. Ree, K. et al.: Polymer *18*, 308 (1977)
152. Norrel, C. J. et al.: J. Polymer Sci., Polymer Chem. Ed. *12*, 913 (1974)
153. Cherkashina, L. G., Berlin, A. A.: Vysokomol. Soedin. *8*, 627 (1966)
154. Bannehr, R., Meyer, G., Wöhrle, D.: Polym. Bull. *2*, 841 (1980)
155. Inoue, H., Kida, Y., Imoto, E.: Bull. Chem. Soc. Jap. *40*, 184 (1967)
156. Epstein, A., Wildi, B. S.: J. Chem. Phys. *32*, 324 (1960)
157. Berlin, A. A. et al.: Vysokomol. Soedin. *6*, 832 (1964)
158. Cherkashina, L. G. et al.: Vysokomol. Soedin. *7*, 1264 (1965)
159. Hanke, W.: Z. anorg. allgem. Chem., *347*, 67 (1966)
160. Hanke, W.: Z. Chem. *6*, 69 (1966)
161. Nasirdinov, S. D. et al.: Zh. Fiz. Khim. *40*, 2614 (1966)
162. Boston, D. R., Bailar, J. C.: Inorg. Chem. *11*, 1578 (1972)
163. Wildi, B. S., Katon, J. E.: J. Polym. Sci. A *2*, 4709 (1964)
164. Sherle, A. I. et al.: Vysokomol. Soedin. A *22*, 1258 (1980)
165. Bellido, J., Cardoso, J., Akachi, T.: Makromol. Chem. *182*, 713 (1981)
166. Bannehr, R., Jaeger, N., Meyer, G., Wöhrle, D.: Makromol. Chem. *12*, 2633 (1981)
167. Wöhrle, D., Meyer, G., Wahl, B.: Makromol. Chem. *181*, 2127 (1980)

168. Liepens, R.: Makromol. Chem. *118*, 36 (1968)
169. Liepens, R., Campbell, D., Walter, C.: A.C.S.M. Polymer Preprints *9*, 765 (1968)
170. Manecke, G., Wöhrle, D.: Makromol. Chem. *120*, 176 (1968)
171. Anderson, D. R., Holovka, J. M.: J. Polym. Sci. A *4*, 1689 (1966)
172. Komarova, O. P. et al.: Vysokomol. Soedin. A *9*, 336 (1967)
173. Marose, U., Meyer, G., Voß, R., Wöhrle, D.: Makromol. Chem., in preparation
174. Wöhrle, D., Bannehr, R., Jaeger, N., Schumann, B.: Makromol. Chem., in press
175. Wöhrle, D., Meyer, G., Voss, R.: Makromol. Chem., in preparation
176. Marvel, C. S., Martin, M. M.: J. Amer. Chem. Soc. *80*, 6600 (1958)
177. Koßmehl, G., Rohde, M.: Makromol. Chem. *178*, 715 (1977)
178a. Griffith, J. R. et al.: Amer. Chem. Div. Org. Coatings Plast. Chem. Pap. *37*, 180 (1977);
178b. – *39*, 546 (1978);
178c. – *40*, 781 (1979);
178d. – *43*, 804 (1980)
179. Ting, R. Y. et al.: Polym. Prep. Amer. Chem. Soc. Div. Polym. Chem. *22*, 50 (1981)
180. Bascom, W. D., Cottington, R. L., Ting, T. Y.: J. Mater. Sci. *15*, 2097 (1980)
181. Murullo, N. P., Snow, A. W.: Polym. Prep. Amer. Chem. Soc. Div. Polym. Chem. *22*, 48 (1981)
182. Marvel, C. S., Rassweiler, J. H.: J. Amer. Chem. Soc. *80*, 1197 (1958)
183. Balabanov, Ye. I., Frankevich, Ye. L., Cherkashina, L. G.: Vysokomol. Soedin. *5*, 1684 (1962)
184. Berezin, B. D., Shormanova, L. P.: Vysokomol. Soedin. A *10*, 384 (1968)
185. Drinkhard, W. C., Bailar, J. C.: J. Amer. Chem. Soc. *8*, 4795 (1959)
186. Berezin, B. D. et al.: Izv. Vyssh. Uchebn. Zaved. Khim. Khim. Tekhnol. *18*, 932, 1446 (1975)
187. Shormanova, L. P., Kolfman, O. I., Berezin, B. D.: Vysokomol. Soedin. B *15*, 910 (1973)
188. Shormanova, L. P., Kojiman, O. I., Berezin, B. D.: Vysokomol. Soedin. B *15*, 910 (1973)
189. Nalwa, H. S., Sinka, J. M., Vasudevan, P.: Makromol. Chem. *182*, 811 (1981)
190. Kreja, L., Plewka, A.: Electrochim. Acta *25*, 1283 (1980)
191. Shormanova, L. P., Berezin, B. D.: Vysokomol. Soedin. A *10*, 1154 (1968)
192. Berezin, B. D., Shormanova, L. P. et al.: Izv. Vyssh. Uchebn. Zaved., Khim. Khim. Tekhnol. *16*, 442 (1973)
193. Blomquist, J. et al.: Inorg. Chim. Acta *53*, L 39 (1981)
194. Meier, H. et al.: J. Phys. Chem. *81*, 712 (1977)
195. Berezin, B. D., Shlyapova, A. N.: Vysokomol. Soedin. A *15*, 1671 (1973)
196a. Akopov, A. S., Lomova, T. N., Berezin, B. D.: Izv. Vyssh. Uchebn. Zaved., Khim. Khim. Tekhnol. *19*, 1177 (1976);
196b. – *21*, 663 (1978)
197. Zhubanov, B. A., Zharmagambetov, A. K.: Izv. Akad. Nauk SSSR, Ser. Khim. *23*, 71 (1973)
198a. Shirai, H. et al.: Makromol. Chem. *178*, 1889 (1977);
198b. – *180*, 2073 (1979)
199. Achar, B. N., Fohlen, G. M., Parker, J. A.: Org. Coat. Plast. Chem. *45*, 31 (1981)
200. Manecke, G., Wöhrle, D.: Makromol. Chem. *102*, 1 (1967)
201. Berlin, A. A. et al.: Vysokomol. Soedin. *4*, 860 (1962)
202. Berezin, B. D., Shormanova, L. P.: Vysokomol. Soedin. A *11*, 1033 (1969)
203. Berlin, A. A., Belova, G. V., Sherle, A. I.: Vysokomol. Soedin., A *14*, 1970 (1972)
204. Schramm, C. J. et al.: J. Amer. Chem. Soc. *102*, 6702 (1980)
205. Frankevich, Ye. L. et al.: Vysokomol. Soedin. *6*, 1028 (1964)
206. Nose, Y. et al.: J. Chem. Soc. Japan, Ind. Chem. Sec. *67*, 1600 (1964)
207. Felsmeyer, W., Wolff, I.: J. Electrochem. Soc. *105*, 141 (1958)
208. Berlin, A. A., Cherkashina, L. G., Balabanov, E. I.: Vysokomol. Soedin. *4*, 376 (1962)
209. Meier, H. et al.: Ber. Bunsenges. Phys. Chem. *77*, 843 (1973)
210. Berlin, A. A., Matveeva, N. G.: Dokl. Akad. Nauk. S.S.S.R. *140*, 368 (1961)
211. Berlin, A. A. et al.: Vysokomol. Soedin. A *10*, 1013 (1970)
212. Rougeè, M., Andre, C.: J. Polymer Sci. C *16*, 3167 (1968)
213. Berlin, A. A. et al.: Dokl. Akad. Nauk, S.S.S.R. *136*, 1127 (1961)
214. Levina, S. D. et al.: Dokl. Akad. Nauk, S.S.S.R. *145*, 602 (1962)
215. Nose, Y. et al.: J. Chem. Soc. Japan, Ind. Chem. Soc. *67*, 1604 (1964)

216. Naraba, T. et al.: Japan J. Appl. Phys. *4*, 977 (1965)
217a. Boguslavskh, L. I., Stil'bans, L. S.: Vysokomol. Soedin. *10*, 1802 (1964)
217b. – Dokl. Akad., Nauk S.S.S.R. *147*, 1114 (1962)
218. Wöhrle, D., Manecke, G.: Makromol. Chem. *120*, 176 (1968)
219. Meier, H., Wöhrle, D.: unpublished
220a. Inoue, H., Kida, Y., Imoto, E.: Bull. Chem. Soc., Japan, *40*, 184 (1967);
220b. – *41*, 684 (1968);
220c. – *41*, 692 (1968)
221. Baker, D. J., Boston, D. R., Bailar, J. C.: D. Inorg. nucl. Chem. *35*, 153 (1973)
222. Hara, T., Ohkatsu, Y., Osa, T.: Chem. Lett. *1973*, 953
223a. Ohkatsu, Y., Hara, T., Osa, T.: Bull. Chem. Soc. Japan, *50*, 696 (1977);
223b. – 701 (1977)
224a. Hara, T., Ohkatsu, Y., Osa, T.: Chem. Lett. *1973*, 103;
224b. –, Bull. Chem. Soc. Japan *48*, 85 (1975)
225. Roginskii, S. Z. et al.: Dokl. Akad. Nauk S.S.S.R. *148*, 118 (1963)
226. Kaneko, M., Manecke, G.: Makromol. Chem. *175*, 2795 (1974)
227. Kaneko, M., Manecke, G.: Makromol. Chem. *175*, 2811 (1974)
228. Kaneko, M., Manecke, G.: Makromol. Chem. *175*, 3401 (1974)
229. Dietrich, H., Storck, W., Manecke, G.: Makromol. Chem. *182*, 2371 (1981)
230. Naito, S., Tamari, K.: Z. Phys. Chem. *94*, 150 (1975)
231. Inoue, H., Aoki, R., Imoto, E.: Chem. Lett. *1974*, 1157
232. Inoue, H. et al.: Bull. Chem. Soc. Jap. *52*, 469 (1979)
233. Takamiya, N. et al.: Nippon Kagaku Kaishi *1977*, 1775
234. Takamiya, N. et al.: Nippon Kagaku Kaishi *1978*, 1078
235. Takamiya, N. et al.: Nippon Kagaku Kaishi *1979*, 825
236. Takamiya, N. et al.: Nippon Kagaku Kaishi *1979*, 1141
237. Roginskii, S. Z. et al.: Kinetika i Kataliz *4*, 431 (1963)
238. Tarasevich, M. R., Zakharkin, G. I.: React. Kinet. Catal. Lett. *6*, 77 (1977)
239. Shirai, H. et al.: Makromol. Chem. *181*, 565 (1980)
240. Kreier, N. D. et al.: Kinetika i Kataliz *2*, 509 (1961)
241. Hanke, W.: Z. Anorg. Allg. Chem. *347*, 67 (1966); *355*, 160 (1967)
242. Acres, G. J. K., Eley, D. D.: Trans. Faraday Soc. *60*, 1157 (1964)
243. Jahnke, H.: Chimia *34*, 58 (1980)
244. Jahnke, H., Schönborn, M., Zimmermann, G.: Topics Current Chem. *61*, 133 (1976)
245. Kretzschmar, C., Wiesener, K.: Z. phys. Chem. *257*, 39 (1976)
246. Musilova, M., Mrha, J., Jindra, J.: J. Appl. Electrochem. *3*, 213 (1973)
247. Johansson, L. Y., Mrha, J., Larsson, R.: Electrochim. Acta *18*, 255 (1973)
248. Kreja, L., Plewka, A.: Electrochim. Acta *25*, 1283 (1980)
249. Appleby, A. J., Savy, M.: Electrochim. Acta *21*, 567 (1976)
250. Appleby, A. J., Fleisch, J., Savy, M.: J. Cat. *44*, 281 (1976)
251. Behret, H., Binder, H., Sandstede, G., Scherer, G. G.: J. Electroanal. Chem. *117*, 29 (1981)
252. Savy, M. et al.: Electrochim. Acta *18*, 191 (1973)
253. Savy, M., Andro, P., Bernard, C.: Electrochim. Acta *19*, 403 (1974)
254. Appleby, A. J., Savy, M.: Electrochim. Acta *22*, 1315 (1977)
255. Maroie, S., Savy, M., Verbist, J. J.: Inorg. Chem. *18*, 2560 (1979)
256. Beck, F. et al.: Z. Naturf. *28a*, 1009 (1973)
257. Beck, F.: Ber. Bunsenges. Phys. Chem. *77*, 353 (1973)
258. Bannehr, R., Jaeger, N., Wöhrle, D.: J. Electrochem. Soc., in preparation
259. Vinogradowa, S. V. et al.: Makromol. *177*, 1905 (1976)
260. Korshak, V. V. et al.: Dokl. Akad. Nauk, SSSR *195*, 1113 (1970)
261. Korshak, V. V., Vinogradowa, S. V. et al.: Faserf. Textiltechnik *26*, 318 (1975)
262. Korshak, V. V., Vinogradowa, S. V. et al.: Dokl. Akad. Nauk SSSR, Ser. Khim. *222*, 114 (1975)
263. Gaymans, R. J., Hodd, K. A., Holmes-Walker. W. A.: Polymer *12*, 400 (1971)
264. Manecke, G., Wöhrle, D.: Makromol. Chem. *120*, 192 (1968)
265. Koßmehl, G., Rohde, M.: Makromol. Chem. *180*, 345 (1979)
266. Packham, D. I., Rackley, F. A.: Polymer *10*, 559 (1969); Chem. and Ind. *1967*, 1254

267. Packham, D. I., Haydon, J. C.: Polymer *11*, 385 (1970)
268. Packham, D. I., Davies, J. D., Racklay, F. A.: Polymer *11*, 533 (1970)
269. Packham, D. I., Davies, J. D., Paisley, H. M.: Polymer *10*, 923 (1969)
270. Korshak, V. V. et al.: Vysokomol. Soyed. *A 23*, 1120 (1981)
271. Siling, S. A., Ponomarev, I. I.: Izv. Akad. Nauk. SSSR, Ser. Khim. *8*, 1871 (1978)
272. Vinogradova, S. V. et al.: Vysokomol. Soedin., *A 21*, 138 (1979)
273. Vinogradova, S. V. et al.: Vysokomol. Soedin., *A 21*, 288 (1979)
274. Gaymans, R. J., Hodd, K. A., Holmes-Walker, W. A.: Polymer *12*, 602 (1971)
275. Korshak, V. V. et al.: Vysokomol. Soedin. *A 14*, 701 (1972)
276. Graham, J., Packham, D. I.: Polymer *10*, 645 (1969)
277. Müller, R., Wöhrle, D.: Makromol. Chem. *176*, 2775 (1975)
278. Müller, R., Wöhrle, D.: Makromol. Chem. *177*, 2241 (1976)
279. Müller, R., Wöhrle, D.: Makromol. Chem. *179*, 2161 (1976)
280. Jones, M. E. B., Thorton, D. A., Webb, R. F.: Makromol. Chem. *49*, 69 (1961)
281. Yukimasa, H. et al.: Makromol. Chem. *178*, 941 (1977)
282. Yukimasa, H., Sawai, H., Takizawa, T.: Makromol. Chem. *179*, 531 (1978); *180*, 1681 (1979)
283 a. Yukimasa, H., Sawai, J.: Makromol. Chem., Rapid. Chem. *1*, 579 (1980);
283 b. –: Makromol. Chem. *182*, 1385 (1981)
284 a. Kim, S. J., Takizawa, T.: Makromol. Chem. *175*, 125 (1974);
284 b. –: Bull. Chem. Soc. Jap. *48*, 2197 (1975);
284 c. –: Makromol. Chem. *176*, 891, 1217 (1975)
285 a. Marvel, C. S. et al.: J. Am. Chem. Soc. *79*, 6000 (1957);
285 b. –: *80*, 832 (1958);
285 c. –: *81*, 2668 (1959)
286. Kuhn, R., Staab, H. A.: Chem. Ber. *87*, 272 (1954)
287. Goodwin, H. A., Bailar, J. C.: J. Am. Chem. Soc. *83*, 2467 (1961)
288. Manecke, G., Wille, R.: Makromol. Chem. *133*, 61 (1970); *160*, 111 (1972)
289. Sawodny, W., Riederer, M.: Angew. Chem. *89*, 897 (1977)
290. Riederer, M., Urban, E., Sawodny, W.: Angew. Chem. *89*, 898 (1977)
291. Riederer, M., Sawodny, W.: Angew. Chem. *90*, 642 (1978)
292. Riederer, M., Sawodny, W.: J. Chem. Res. (S), 1978, 450
293. Sawodny, W., Riederer, M., Urban, E.: Inorg. Chim. Acta *29*, 63 (1978)
294. Bottino, F. A. et al.: Inorg. Nucl. Chem. Letters *16*, 417 (1980)
295. Patel, M. N., Patil, S. H., Setty, M. S.: Angew. Makromol. Chem. *97*, 69 (1981)
296 a. Maurya, P. L., Agarwala, B. v., Dey, A. K.: Polym. Bull. *3*, 253 (1980);
296 b. –: Makromol. Chem. *183*, 511 (1982)
297 a. Karampurwala, A. M. et al.: Makromol. Chem. *181*, 57 (1980);
297 b. –: Makromol. Sci. Chem. *A 15*, 431, 439 (1981)
298. Rana, A. K. et al.: Makromol. Chem. *182*, 3387 (1981)
299. Wille, F., Schwab, W.: Monatshefte Chem. *108*, 929 (1977)
300. Spiratos, M., Airinei, A., Ciobanu, A.: Rev. Roumaine Chim. *25*, 1083 (1980)
301. Bhaduri, S., Khwaja, H., Khanwalkar, V.: J. Chem. Soc., Dalton Trans. *1982*, 445
302 a. Loutfy, R. O. et al.: J. Chem. Phys. *71*, 1211 (1979);
302 b. –: J. Appl. Phys. *52*, 5218 (1981)

Received September 9, 1982
H.-J. Cantow (Editor)

Subject Index Volumes 25–50

The volume numbers are printed in italics

Author Index Volumes 1–50

Cerf, R.: La dynamique des solutions de macromolecules dans un champ de vitesses. Vol. 1, pp. 382–450.

Cesca, S., Priola, A. and *Bruzzone, M.:* Synthesis and Modification of Polymers Containing a System of Conjugated Double Bonds. Vol. 32, pp. 1–67.

Cicchetti, O.: Mechanisms of Oxidative Photodegradation and of UV Stabilization of Polyolefins. Vol. 7, pp. 70–112.

Clark, D. T.: ESCA Applied to Polymers. Vol. 24, pp. 125–188.

Coleman, Jr., L. E. and *Meinhardt, N. A.:* Polymerization Reactions of Vinyl Ketones. Vol. 1, pp. 159–179.

Crescenzi, V.: Some Recent Studies of Polyelectrolyte Solutions. Vol. 5, pp. 358–386.

Davydov, B. E. and *Krentsel, B. A.:* Progress in the Chemistry of Polyconjugated Systems. Vol. 25, pp. 1–46.

Dole, M.: Calorimetric Studies of States and Transitions in Solid High Polymers. Vol. 2, pp. 221–274.

Dreyfuss, P. and *Dreyfuss, M. P.:* Polytetrahydrofuran. Vol. 4, pp. 528–590.

Dušek, K. and *Prins, W.:* Structure and Elasticity of Non-Crystalline Polymer Networks. Vol. 6, pp. 1–102.

Eastham, A. M.: Some Aspects of the Polymerization of Cyclic Ethers. Vol. 2, pp. 18–50.

Ehrlich, P. and *Mortimer, G. A.:* Fundamentals of the Free-Radical Polymerization of Ethylene. Vol. 7, pp. 386–448.

Eisenberg, A.: Ionic Forces in Polymers. Vol. 5, pp. 59–112.

Elias, H.-G., Bareiss, R. und *Watterson, J. G.:* Mittelwerte des Molekulargewichts und anderer Eigenschaften. Vol. 11, pp. 111–204.

Elyashevich, G. K.: Thermodynamics and Kinetics of Orientational Crystallization of Flexible-Chain Polymers. Vol. 43, pp. 207–246.

Fischer, H.: Freie Radikale während der Polymerisation, nachgewiesen und identifiziert durch Elektronenspinresonanz. Vol. 5, pp. 463–530.

Fradet, A. and *Maréchal, E.:* Kinetics and Mechanisms of Polyesterifications. I. Reactions of Diols with Diacids. Vol. 43, pp. 51–144.

Fujita, H.: Diffusion in Polymer-Diluent Systems. Vol. 3, pp. 1–47.

Funke, W.: Über die Strukturaufklärung vernetzter Makromoleküle, insbesondere vernetzter Polyesterharze, mit chemischen Methoden. Vol. 4, pp. 157–235.

Gal'braikh, L. S. and *Rogovin, Z. A.:* Chemical Transformations of Cellulose. Vol. 14, pp. 87–130.

Gallot, B. R. M.: Preparation and Study of Block Copolymers with Ordered Structures, Vol. 29, pp. 85–156.

Gandini, A.: The Behaviour of Furan Derivatives in Polymerization Reactions. Vol. 25, pp. 47–96.

Gandini, A. and *Cheradame, H.:* Cationic Polymerization. Initiation with Alkenyl Monomers. Vol. 34/35, pp. 1–289.

Geckeler, K., Pillai, V. N. R., and *Mutter, M.:* Applications of Soluble Polymeric Supports. Vol. 39, pp. 65–94.

Gerrens, H.: Kinetik der Emulsionspolymerisation. Vol. 1, pp. 234–328.

Ghiggino, K. P., Roberts, A. J. and *Phillips, D.:* Time-Resolved Fluorescence Techniques in Polymer and Biopolymer Studies. Vol. 40, pp. 69–167.

Goethals, E. J.: The Formation of Cyclic Oligomers in the Cationic Polymerization of Heterocycles. Vol. 23, pp. 103–130.

Graessley, W. W.: The Etanglement Concept in Polymer Rheology. Vol. 16, pp. 1–179.

Graessley, W. W.: Entagled Linear, Branched and Network Polymer Systems. Molecular Theories. Vol. 47, pp. 67–117.

Hagihara, N., Sonogashira, K. and *Takahashi, S.:* Linear Polymers Containing Transition Metals in the Main Chain. Vol. 41, pp. 149–179.

Hasegawa, M.: Four-Center Photopolymerization in the Crystalline State. Vol. 42, pp. 1–49.

Hay, A. S.: Aromatic Polyethers. Vol. 4, pp. 496–527.

Hayakawa, R. and *Wada, Y.:* Piezoelectricity and Related Properties of Polymer Films. Vol. 11, pp. 1–55.

Heidemann, E. and *Roth, W.:* Synthesis and Investigation of Collagen Model Peptides. Vol. 43, pp. 145–205.

Heitz, W.: Polymeric Reagents. Polymer Design, Scope, and Limitations. Vol. 23, pp. 1–23.

Advances in Polymer Science

Fortschritte der
Hochpolymeren-Forschung

Editors:
H.-J. Cantow, G. Dall'Asta,
K. Dušek, J. D. Ferry, H. Fujita,
M. Gordon, J. P. Kennedy,
W. Kern, S. Okamura,
C. G. Overberger, T. Saegusa,
G. V. Schulz, W. P. Slichter,
J. K. Stille

Springer-Verlag
Berlin
Heidelberg
New York

Volume 41

Speciality Polymers

1981. 80 figures. V, 186 pages. ISBN 3-540-10554-9

Contents/Information:

K. Takemoto, Y. Inaki: **Synthetic Nucleic Acid Analogs: Preparation and Interactions.** The functional monomers and polymers containing heterocyclic moieties have recently received much attention. Numerous studies have been devoted to the preparation and polymerization of these new monomeric species, which may find a number of application possibilities as polymeric drugs and other biomaterials. This article focusses on the authors work on specific base – base interactions between the nucleic acid analogues. (85 references)

A. Y. Grosberg, A. R. Khokhlov: **Statistical Theory of Polymeric Lyotropic Liquid Crystals.** This article deals with following topics of the statistical physics of liquid-crystalline phase in the solutions of stiff chain macromolecules: problems of the phase diagram for the liquid-crystalline transition in the solutions of completely stiff macromolecules (rigid rods); conditions of formation of the liquid-crystalline phase in the solutions of semiflexible macromolecules; possibility of the intramolecular liquid-crystalline ordering in semiflexible macromolecules; structure of intramolecular liquid crystals and dependence of the properties of the liquid-crystalline phase on the microstructure of the polymer chain. (45 references)

E. A. Bekturov, L. A. Bimendina,: **Interpolymer Complexes.** The problems of complex formation in different systems of interacting macromolecules, in polymer-polymer, polymer-alternating or statistical copolymer systems are discussed. The influence of solvent nature, the critical phenomena, equilibrium, selectivity and cooperativity in reactions are considered. The perspectives of development of this field of polymer science and the potential practical applications of interpolymer complexes are pointed out. (179 references)

N. Hagihara, K. Sonogashira, S. Takahashi: **Linear Polymers Containing Transition Metals in the Main Chain.** Recent studies on linear polymers containing transition metals in the main chain are reviewed on the basis of metallocene-containing polymers, linear "Werner Type" coordination polymers and polymers containing sigmabonded transition metals in the main chain. (56 references)

Advances in Polymer Science

Fortschritte der Hochpolymeren-Forschung

Editors:
H.-J. Cantow, G. Dall'Asta, J. D. Ferry, H. Fujita,
W. Kern, G. Natta, S. Okamura, C. G. Overberger,
W. Prins, G. V. Schulz, W. P. Slichter, A. J. Staverman,
J. K. Stille, H. A. Stuart

Volume 10

1972. 50 figures. III, 194 pages. ISBN 3-540-05838-9

Contents:

Springer-Verlag
Berlin
Heidelberg
New York